An Introduction to Ultracentrifugation

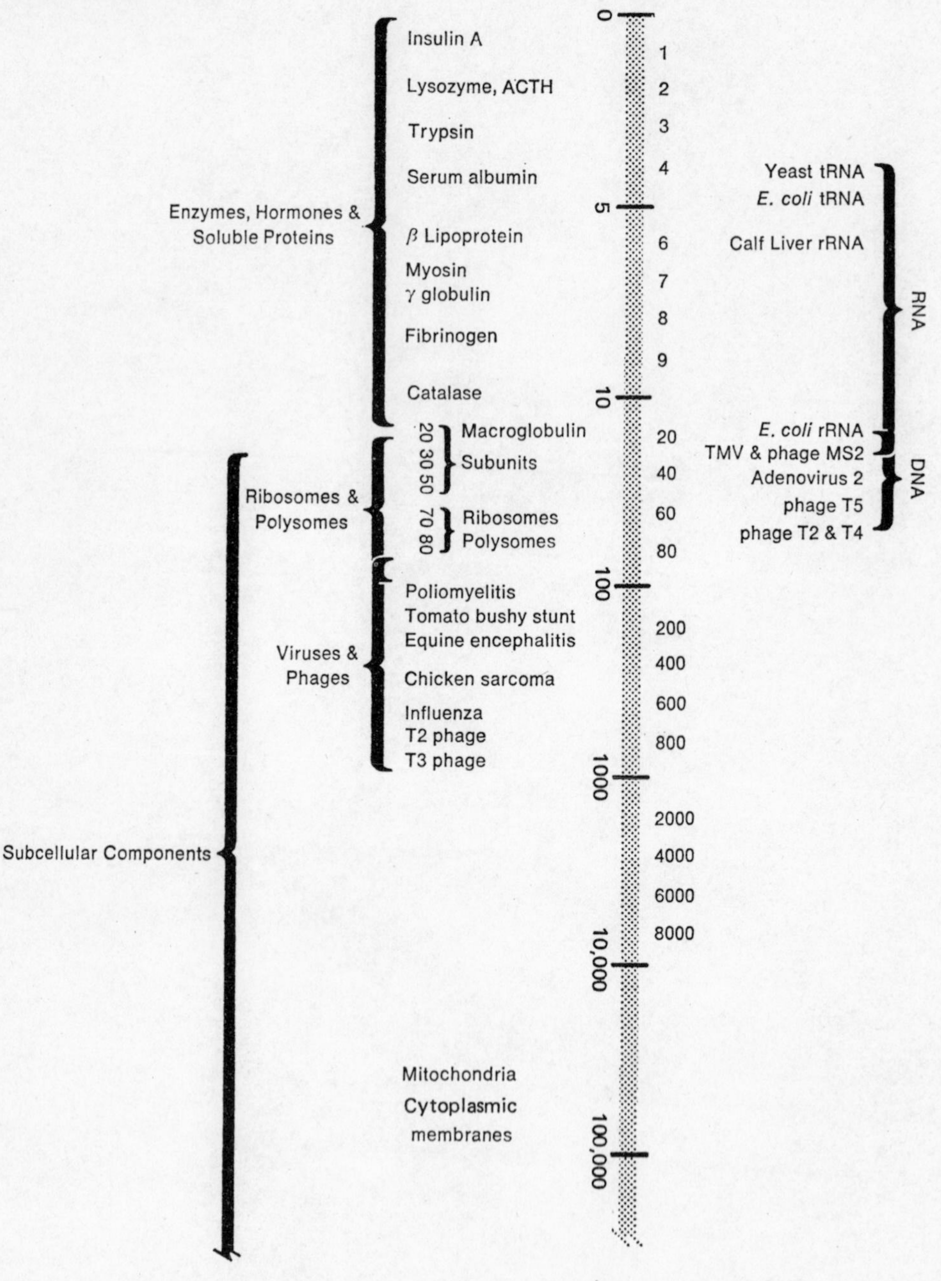

Frontispiece (*Courtesy of Beckman Instruments Inc.*)

An Introduction to Ultracentrifugation

T. J. Bowen

Senior Lecturer in Biochemistry,
The University of Leeds

With additional material by

A. J. Rowe

Department of Biochemistry,
University of Leicester

WILEY-INTERSCIENCE

a division of John Wiley & Sons Ltd.

LONDON NEW YORK SYDNEY TORONTO

Library of Congress Catalog card No. 79-129158

ISBN 0 471 09215 0

Reprinted 1971

Printed in Great Britain by
Unwin Brothers Limited
The Gresham Press, Old Woking, Surrey, England
A member of the Staples Printing Group

TO VERA

Foreword

In the very first article to appear in that invaluable series, *Advances in Enzymology*, H. B. Bull described the ultracentrifuge as 'the most important tool ever devised for the physical study of proteins'. More than a quarter of a century has elapsed since these words were written, and many techniques of great value have been devised during this period, but the ultracentrifuge has continued to occupy a central position amongst the methods available for investigating biological macromolecules. Most educated people, whether they are scientists or not, recognize that the X-ray measurements made on deoxyribonucleate by M. H. F. Wilkins, and their interpretation by J. D. Watson and F. H. C. Crick in 1953, initiated a revolution of thought in the biological sciences. The tremendous increase in our knowledge of those macromolecules that determine inheritance, and of the events that occur when proteins are synthesized by cells, also owes a great deal to the many ingenious applications of ultracentrifugation used by numerous investigators.

Thus, proof that the strands of a double helix of DNA untwined and separated during cell division, without undergoing degradation, was provided in 1958 by the experiments of M. Meselson and F. W. Stahl. *Escherichia coli* which had previously incorporated ^{15}N into DNA and other cellular materials was allowed to divide in fresh medium containing ^{14}N, and in consequence the bacterial DNA decreased in density. This decrease was followed for several generations by observing three separate bands of DNA which formed in caesium chloride solution as equilibrium was closely approached at 140,000 g in the analytical ultracentrifuge. This instrument was also used to demonstrate the existence of bacterial ribosomes several years before these particles had been named and their function in metabolism suggested. In 1951, H. K. Schachman, A. B. Pardee and R. Y. Stanier found that cell-free extracts from various bacterial species all contained ribonucleoprotein components that sedimented at about 40 S and 19 S (uncorrected); these were the 50 S and 30 S ribosomes which were given their present names at the first symposium of the Biophysical Society held in Boston in 1958. During this same investigation, the chromatophores of photosynthetic bacteria were also discovered and named.

The issue of *Nature* of May 13th, 1961, contained contributions by two groups of workers that proved to be of singular importance for our present understanding of the process of protein biosynthesis. Both investigations employed the preparative ultracentrifuge; in one, the constituents of cell-free extracts were partially separated in density gradients formed by centrifuging caesium chloride solutions, and in the other, sucrose density gradients were used. The first communication (Brenner, Jacob and Meselson) showed that when *Escherichia coli* was infected by bacteriophage, the 'old' ribosomes of the host cells were used during the synthesis of new phage protein. The second set of experiments (Gras and coworkers) used pulse-labelling of uninfected cells to demonstrate the existence of a transitory 'messenger' RNA which appeared to attach itself to 70 S ribosomes. Together, these papers supported the suggestion of F. Jacob and J. Monod which had recently been made, that ribosomes are non-specialized structures that receive genetic information from messenger RNA. Later work with the ultracentrifuge has supported the evidence of electron microscopy, that a number of 70 S ribosomes may attach to a strand of messenger RNA to form a polysome. If such strands remain intact when the cells are broken, boundaries may be observed that sediment at speeds predicted for structures containing several ribosomes.

Accordingly, we find that the ultracentrifuge is now used continuously in a large number of laboratories where interests are centred upon nucleic acids and protein biosynthesis. The instrument is also indispensable in a related area of increasing research activity, namely the study of mechanisms by which enzymic reactions are regulated. Rates of reaction sequences are adjusted to meet the metabolic requirements of cells mainly through the responses made by allosteric enzymes to changes in the concentrations of individual metabolites. Such responses, in turn, depend upon changes in conformation which these metabolites can bring about, and which are apparently transmitted through interactions between the subunits that constitute the complete enzyme molecule. The fruitful attempts to characterize the allosteric enzyme aspartate transcarbamylase illustrate the value of studies with the ultracentrifuge, especially when they are combined with other methods of analysis. For example, J. C. Gerhart and H. K. Schachman (1965) showed that this enzyme, a globular protein of molecular weight about 3×10^5, gave rise to sub-units of two different types when treated with mercurials. One type catalysed the reaction and the other, which did not, contained a site of attachment for cytidine triphosphate; thus 5-bromocytidine triphosphate sedimented with this non-catalytic, or regulatory, sub-unit. Later K. Weber (1968) determined the amino acid compositions and amino terminal residues of the two types and showed that one molecule of the enzyme contained four regulatory and four catalytic polypeptide chains. The earlier work had indicated only two catalytic particles per molecule, so that the catalytic sub-units presumably sedimented as dimers. It is evident that once the ultracentrifuge has shown the existence of different types of sub-units in a complex enzyme, other techniques such as X-ray diffraction are required to reveal their arrangement in space. A further limitation of the scope of the ultracentrifuge in this area lies in the present uncertainty as to the values of partial specific volumes of proteins in solvents used to dissociate molecular complexes into sub-units.

It always provides great satisfaction to a biochemist when he is able to crystallize an enzyme under investigation, but he will also be aware that this pleasant event affords no guarantee of complete purity. W. B. Jakoby (1968), in whose laboratory fifty proteins have been induced to crystallize in three years, has also reported the crystallization of preparations at 30 per cent of purity; he has re-emphasized the need for applying criteria of homogeneity, such as those provided by ultracentrifugation and electrophoresis. On the other hand, a heterogeneous mixture may appear to give rise to a single peak in the ultracentrifuge if the experiment be performed in an uncritical manner. The speeds of particles as they sediment in a centrifugal field, and the characteristics of the peaks to which they give rise, are influenced by many factors that are discussed in this book and which any investigator ignores at his peril. This is one reason why, despite the power of this technique in the areas briefly mentioned, it is found that the rules of many laboratories restrict the operation of the ultracentrifuge to persons who have been thoroughly instructed in the use of the instrument and the interpretation of the data it provides. Those who already possess this knowledge generally receive a warm welcome to research teams and this book will stimulate the supply of such valuable recruits. The treatment of the subject is sound but not over-sophisticated, for it is the result of many years of thought devoted by an outstanding teacher to the problems of presenting physicochemical concepts to students of biochemistry.

From time to time former students, who now hold responsible positions in laboratories in Britain and North America, have told me of the debt they owe to Mr. Bowen's instruction which is now available to all. In addition, his present and former teaching colleagues, of whom I am one, will join in expressing their gratitude for unofficial tuition that was given most generously whenever it was sought.

Brenner, S., F. Jacob and M. Meselson (1961) *Nature Lond.*, **190,** 576.
Gerhart, J. C. and H. K. Schachman (1965) *Biochemistry*, **4,** 1054.
Gros, F., H. Hiatt, W. Gilbert, C. G. Kurland, R. W. Risebrough and J. D. Watson (1961) *Nature Lond.*, **190,** 581.
Jakoby, W. B. (1968) *Anal. Biochem.*, **26,** 295.
Meselson, M. and F. W. Stahl (1958) *Proc. natn. Acad. Sci., U.S.A.*, **44,** 671.
Schachman, H. K., A. B. Pardee and R. Y. Stanier (1951) *Arch. Biochem. Biophys.*, **38,** 245.
Weber, K. (1968) *J. Biol. Chem.*, **243,** 543.

Department of Biochemistry,
University of Minnesota,
St. Paul, Minnesota,
U.S.A.

S. DAGLEY

Preface

The ultracentrifuge has been an increasingly important instrument in the field of polymer science for nearly half a century. As a routine tool it serves the biologists for preparative and analytical work, since very little material is needed for the determination of a considerable number of parameters and separations of great delicacy are made possible. The needs of users vary widely; there are biophysicists interested in the subtleties of interacting systems and in the application of computer technology to the solution of mathematical relationships once considered to be too tedious to be solved. On the other hand, many biologists require a quick answer to the question of purity and size of an active preparation that they have just isolated. The object of this book is to serve as an introduction to the subject of ultracentrifugation, eventually leading to a study of the many advanced reviews quoted within. It is hoped that it will be particularly valuable to students and new research workers. Biophysicists will certainly need to read the further articles mentioned, but many biologists may find certain sections more than they may require. In any case, the book should enable the beginner to latch on to a great deal of the discussion encountered in symposia.

The contents of the book are largely derived from a course of lectures given to students of biochemistry (and subsequently to the biophysicists) at the University of Leeds, England. Certain students who expressed a desire for further training were given a practical course on the ultracentrifuge. Most of the students were given photographic plates obtained using the machine and instructed in methods of working out the data. Recently, the shortening of the Leeds course by one year to bring it into conformity with courses at other British universities made it appear imperative to write a book based on the original course. Students had long expressed a wish to read something more detailed than isolated one-chapter surveys of parts of the subject in general texts on methodology. On the other hand, the same students did not feel ready to plunge immediately into the advanced books and reports of meetings; it is hoped that this book will prove useful in this sense. The mathematical content has been kept to a minimum but is nowhere exacting. The Leeds course was developed over twenty years, so the references quoted for each chapter represent those found to be particularly helpful to the author; perhaps others will find them as valuable.

Chapter 6 on the determination of molecular weights owes much to Mr. Peter Ashby who, as a final year student under the original course, tried various techniques as a research problem. It is possible that repetition of experiments could have refined the data but the reader may find them a useful guide to what a complete beginner on the ultracentrifuge was able to achieve. Mr. Ashby proved to be above average in his practical ability and scientific drive, and the author would like to express his gratitude to him for his enthusiastic and skilled support. Credit is also due to my colleague, Dr. J. H. Parish, who patiently read the whole of the first draft and eliminated many mistakes. Dr. A. J. Rowe of the University of Leicester read the book in its preliminary stages and made many helpful criticisms. Any mistakes still remaining must certainly be attributed to the author. Dr. Rowe also contributed Chapters 10 and 11 and some illustrations, etc. to improve earlier chapters. Dr. G. A. Gilbert of the Department of Biochemistry in the University of Birmingham kindly gave his permission to use illustrations from his work in Chapter 8. I must also express my thanks to my colleague, Dr. Hassall, for his constant encouragement and his ability to ask awkward questions.

Many illustrations of equipment were provided by manufacturers; their names appear within where their photographs and diagrams have been used. Their help was promptly given and is much appreciated.

Department of Biochemistry,
University of Leeds,
England.

T. J. Bowen

Main Symbols Used

Symbol	Meaning	Page
A	Area of plane	39
	Area of Schlieren peak	44, 47
a	Optical depth	30
	Half-length of rigid polymeric molecule	144
b	Optical lever arm	20
	Half-width of rigid polymeric molecule	144
C	Cylinder lens enlargement factor	20
c	Solute concentration with suitable primes to indicate whether it refers to meniscus, plateau, time, etc	—
D	Translational diffusion coefficient	39
D°_{20},w	D corrected to pure water as solvent at 20°C and extrapolated to zero solute concentration	43
d	Optical depth	17
F	Enlargement factor from cell to photograph	20, 78
	Ratio between f and f_0	91
	Function symbol	—
f	Frictional coefficient	1, 84
f_0	f for a spherical particle	1, 85
g	Earth's gravitational field, 980·6 cm/sec^2	2
H	Maximum height of Schlieren peak:	44
	also as defined	67
h	Optical pathlength	51
I_0,I	Intensity of incident and emergent light	17
j	Number of interference fringe	30
K	Equilibrium constant	110
k	Specific refraction increment	17
	Heavy water binding factor	59
	Constant relating s to c (also k_s)	32
	Measure of efficiency of preparative rotor	128
	$M(1-\bar{v}\rho)\omega^2/RT$	141
k'	Measure of preparative rotor efficiency (zonal)	133
k',k''	Constants (Huggins) in viscosity determination	88
k_{+1},k_{-1}	Rate constants for balanced reaction	109
L	Height of liquid in Anderson rotor	134
M	Molecular weight (daltons)	—
	Concentration of monomer	103
M_n,M_w,M_z	Number, weight and z-average for M	61
m	Mass	51, 80
N	Avogadro's number, $6{\cdot}023\times10^{23}$	3
n	Degree of polymerization	103

n_s, n_0	Refractive indices for solution and solvent	16
P	Polymer concentration	103
P_i	Preparative rotor performance index	129
p, q	Axial ratios of ellipsoids	90
R	Gas constant, $8{\cdot}313 \times 10^7$ erg/mole/°K	4
r	Radius of spherical particle	1
	Binding ratio for small molecules	154
RCF	Relative centrifugal field	2
S	Svedberg unit	3
S^*_{min}	Minimum s rate for sedimentation from meniscus to bottom of tube	128
$\overline{ST}$	Method of expressing efficiency of preparative rotor	128
s	Sedimentation coefficient	3, 31
$s^\circ_{20,w}$	s corrected for water as solvent at 20°C and extrapolated to zero concentration	34
T	Temperature, °K	4
t	Time	—
V_e	Volume (hydrated) of particle	90
v	Velocity	
$\bar{v}$	Partial specific volume (unhydrated)	56
w	Mass of solvent bound to 1 g of solute	86
x	Radial distance	51
α	Interfacial region	101
β	Interfacial region	101
	Scheraga–Mandelkern function	91
γ	Interfacial region	101
	Activity coefficient	41, 60
δ	Gilbert theory value of x/vt	105
	Increment symbol	—
ϵ	Molar extinction coefficient:	17
	also as defined	55
η	Coefficient of viscosity	87
$[\eta]$	Intrinsic viscosity	87
θ	Sector angle of ultracentrifuge cell	51
	Phaseplate angle	20
λ	Wavelength of light:	30
	also as defined	47
μ	Ionic strength, $\frac{1}{2}\Sigma c_i z_i^2$	89
ν	Viscosity increment	90
$\prod$	Osmotic pressure	40
π	Ratio of circumference to diameter, 3·1416	—
ρ	Density of solution	1
ρ_P	Density of particle	1
Σ	Summation symbol	—
τ	Equivalent time of centrifugation	47, 52
	Shearing force	87
Φ	Volume fraction	90
	Function symbol	55
ω	Angular velocity, radians/sec	2

Contents

1 The vintage years 1

The pre-war development of ultracentrifugation . . . 2
Molecular weights 3
The first ultracentrifuges 5
Some early results 6

2 The modern ultracentrifuge 8

Analytical ultracentrifuges 9
The Beckman Model E 9
The MSE ultracentrifuges 9
The Christ ultracentrifuge 11
The Metrimpex ultracentrifuge 12
Preparative ultracentrifuges 12

3 Optical systems 15

The absorption system 17
The Schlieren system 18
Cylindrical lens Schlieren 20
Phaseplate Schlieren 22
Optical alignment 23
The Rayleigh interferometer 23
Special cells 30

4 The importance of s and D 31

The sedimentation coefficient 31
Concentration dependence of s 32

Charge effects 32
Corrections for s 34
The experimental determination of s 35
Band centrifugation 37
Determination of s from density gradients . . . 38
The diffusion coefficient 39
Fick's law of diffusion 39
The experimental determination of D 44
Static methods 44
Determination of D in the ultracentrifuge 46

5 A. The continuity equation and relationships derived from it .
B. Partial specific volume 50

The Lamm equation 50
The plateau region 52
Radial dilution 53
Sedimentation equilibrium 53
Approach-to-equilibrium 54
Solutions of the Lamm equation 55
Density and partial specific volume 56
Method (1) 57
Method (2) 58
Method (3) 58

6 The determination of molecular weights 60

Molecular weight averages 61
Machine presentation of data 62
Sedimentation–diffusion 65
Equilibrium methods 65
The high speed equilibrium method 67
Meniscus depletion of Yphantis method 67
Low speed equilibrium methods 70
Lansing and Kraemer method 70
LaBar's method 71
Methods of van Holde and Baldwin and of Lamm . . . 73
The approach-to-equilibrium (Archibald) method . . . 75
Molecular weight via buoyancy 79

7 The determination of molecular conformation 84

Sedimentation data and conformation 85
The equations 85
Effect of asymmetry and hydration 86
Viscosity data and conformation 87
The equations 87
The effect of salts on viscosity for polyelectrodes 88
Conformation models 89
Viscosity data related to conformation 89
The approach to conformation using sedimentation and viscosity data 90
The β function of Scheraga and Mandelkern 90
Conformation and the Wales–van Holde ratio 93
Determination of intrinsic viscosity 94

8 Quantitative analysis of mixtures 97

A. Instrumental factors 97
B. Properties of the system itself 98
The relationship between solute concentration and light absorption or refraction 98
The Johnston–Ogston effect 99
Interacting systems and the Gilbert theory 102
Complexes 109
The problem of gels 112

9 Preparative ultracentrifugation 115

Relationships between parameters 116
Viscosity and density corrections 117
Effect of asymmetry 118
The determination of s from preparative runs 119
Determination of particle density 120
The correlation of activity with a component of a mixture . 120
Classification of methods for centrifugation 121
Rate sedimentation 121
Density gradient separations 121
The importance of rotor and cell design 122
The analytical cell 122

Swing-out and angle heads 123
The Anderson rotor 125
Selection of rotors and methods 127
Non-zonal rate method 127
Zonal methods 130
The capacity of a density gradient 134

10 Some applications of the analytical ultracentrifugation to biological problems* **137**

Criteria of purity 137
Analysis of the sedimenting boundary 138
Analysis of equilibrium experiments 140
The sedimentation of highly asymmetric particles 144
The detection of conformation changes in macromolecules . 146
The determination of chain molecular weights 151
The use of double-beam absorption optics and the 'scanner' . 153

11 Problems dealt with and problems to solve* **158**

Problems dealt with 158
Sedimentation velocity experiments 158
Approach-to-equilibrium experiment 'Archibald method' . . 160
Short-column equilibrium experiments 161
Problems to solve 163
Answers to problems 164

Index 167

* Contributed by A. J. Rowe

CHAPTER 1

The vintage years

The assumption underlying all the work done with an ultracentrifuge is that the parameters of the molecules under investigation may be related to the behaviour of those molecules in a gravitational field. These parameters include molecular weight, shape and density; further parameters are required to account for the effects of the solvent. The assumption is made that the following equation is valid:

$$\text{Force applied to molecule} = f \times \text{velocity of molecule} \tag{1.1}$$

Such a relationship is true only for small velocities; terms involving higher powers of the velocity may be required if velocities are not small. In 1856, Stokes showed that f_0, the frictional coefficient for spherical particles, could be expressed by:

$$f_0 = 6\pi\eta r \tag{1.2}$$

where η is the coefficient of viscosity of the medium and r is the radius of the sphere. Of course, other expressions are needed for the frictional coefficient in terms of molecular geometry for shapes that are not spherical.

If the spheres are acted on by a gravitational field then Equation (1.1) can be written in the form:

$$\tfrac{4}{3}\pi r^3 g(\rho_p - \rho) = 6\pi\eta r \cdot v \tag{1.3}$$

The left hand side of the equation has expressed the force in terms of the volume of the sphere times its density, ρ_p, corrected for buoyancy due to the density of the solvent, ρ, and expressed in dynamic units by the factor, g, the gravitational field. Stokes' equation (1.3) has been much used by physicists to calculate the radius of a spherical droplet when descending under the earth's gravitational field. A classical example of such a use of the equation was by Millikan (1913) when he determined the charge on the electron by the famous oil drop experiment using the Stokes' equation to determine the size of the drops of oil.

Obviously, it was apparent that the size of smaller particles would require greater gravitational fields such as could be produced in a centrifuge:

$$\text{Centrifugal field} = x\omega^2 \tag{1.4}$$

where x is the radius and ω the angular velocity in radians per second; one revolution equals 2π radians. The magnitude of the field has the units of an acceleration, cm/sec^2. The centrifugal field generated in a centrifuge relative to the earth's gravitational field is called, simply, the Relative Centrifugal Field (RCF). From Equation (1.4) it follows that:

$$\mathrm{RCF} = \frac{x\omega^2}{g} \tag{1.5}$$

The pre-war development of ultracentrifugation

Svedberg and Pedersen (1940) have described the early years of the subject leading to the development of very high gravitational fields and their classic book on the subject can still be read with advantage today. The early design work was carried out between the two World Wars but much of the design work on rotors and cells laid down then is still largely valid. This period can be truly described as the heroic phase when engineering science had to be stretched to new limits. Previously, Dumansky (1913) had used an ordinary laboratory centrifuge in an attempt to correlate particle size with observations made using an ultramicroscope; he was not successful because the centrifuge did not provide ideal conditions for convection-free sedimentation. In the early 1920's, Svedberg and his associates pioneered the design of centrifuges that could produce more ideal conditions for sedimentation and its observation than were available to Dumansky. In 1923, a small optical centrifuge was developed to give a field of 150g, direct observation and photography being used to record results. Progress was made in the direction of higher speeds and in the control of temperature and vibration. Although early work was carried out on systems such as gold sols it was apparent that the whole world of the biological colloids was awaiting investigation. The term 'ultracentrifuge' was proposed for a centrifuge with an optical system. The name 'supercentrifuge' (used by the Sharples Company) was reserved for a fast centrifuge without an optical system. However, the distinction has now become blurred and the name ultracentrifuge is now used for preparative or analytical machines in which the rotor runs at high speeds in a vacuum or low pressure hydrogen.

In the early 1920's, the question of the molecular size of protein molecules was still unresolved. Were they large or small, and were they made up of a few discrete sizes or was there a distribution of molecular weights? A minimum molecular weight, analogous to an empirical formula of classical chemistry, was proposed for haemoglobin from the iron content of the molecule, but the actual molecular weight could have been (and is) a multiple of this. The answer came partly from measurements of osmotic pressures (Adair, 1925, 1928) and partly from Svedberg's ultracentrifuge.

The need to develop faster and more refined ultracentrifuges was of pressing urgency.

Molecular weights

There were two main approaches to the determination of molecular weight. The velocity method depended on the application of a sufficiently high gravitational field to cause the sedimentation of the molecule at a rate that could be directly measured. Large viruses could be sedimented at comparatively low speeds, but the smaller molecules required higher gravitational fields. If M is the molecular weight and N is Avogadro's number then we can write, using (1.1) and $\bar{v} = 1/\rho_p$:

$$\frac{M}{N}(1-\bar{v}\rho)x\omega^2 = f \cdot \frac{\mathrm{d}x}{\mathrm{d}t} \tag{1.6}$$

In this equation $\bar{v}$ is the partial specific volume, which can be considered to be the volume occupied by 1 gram of the solute when dissolved. Thus the $\bar{v}\rho$ term allows for the buoyancy effect of the solvent. The buoyancy effect may be clearer if Figure 1.1 is referred to. The weight of each molecule is M/N so the downward force exerted is obtained by multiplying by the gravitational field, $x\omega^2$ as shown in Figure 1.1(a). However, when the partial specific volume is $\bar{v} = 1/\rho_p$ for each gram of dissolved macromolecule it will displace a volume of solvent $(M/N)\bar{v}$ and produce an upthrust as shown in Figure 1.1(b) according to Archimedes' principle. Figure 1.1(c) shows that the combined effects are the algebraic sum of those shown at a and b producing a resultant with the buoyancy correction term, $(1-\bar{v}\rho)$. By definition:

$$s = \frac{\mathrm{d}x/\mathrm{d}t}{x\omega^2}\ \text{sec} \tag{1.7}$$

The sedimentation coefficient, s, is of the order of 10^{-13} seconds so

$$1\ S\ \text{(Svedberg)} = 1\times 10^{-13}\ \text{seconds} \tag{1.8}$$

We can therefore write (1.6) as:

$$\frac{M}{N}(1-\bar{v}\rho) = fs \tag{1.9}$$

If D is the translational diffusion coefficient then (anticipating Chapter 4) we have:

$$D = \frac{RT}{Nf} \tag{1.10}$$

where R is the gas constant, equal to $8{\cdot}313 \times 10^7$ erg/mole/°K, and T is the absolute temperature.

If we combine (1.9) and (1.10) we get:

$$M = \frac{RTs}{D(1-\bar{v}\rho)} \tag{1.11}$$

This was an early derivation of the velocity equation (or Svedberg equation) but more elegant proofs can be proposed and we now know that there is a further thermodynamic term which becomes small at low concentrations

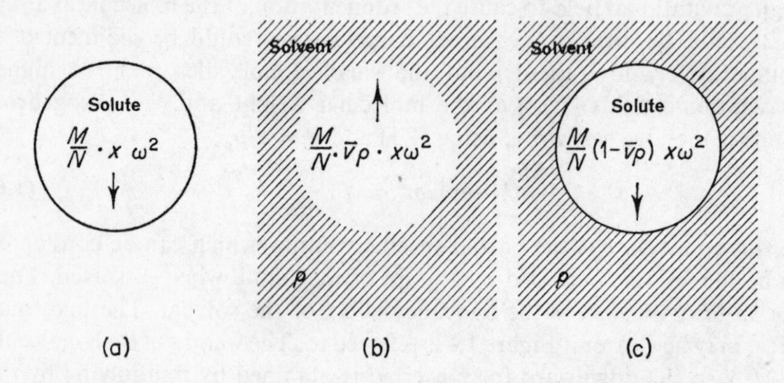

Figure 1.1. The buoyancy correction $(1-\bar{v}\rho)$ arises as a result of the principle of Archimedes. In (a) there is no upthrust from solvent displaced so particle acts with force $(M/N)x\omega^2$; in (b) upthrust is due to displacement of solvent of density ρ, and (c) shows the resultant of (a) and (b)

of solute. The coefficients s and D are dependent on concentration, so they should be expressed at the extrapolated zero concentration, at the same temperature and in the same solvent (usually expressed as the result for water, see Chapter 4).

When pure specimens of the solute were available it was possible to use the sedimentation equilibrium result. In this technique a speed was chosen for the ultracentrifuge at which a balance could be established between sedimentation and diffusion. After a long time, which might be days or weeks using early techniques, a steady state was established. This enabled the molecular weight to be determined by the equilibrium equation (this is derived in Chapter 5):

$$M = \frac{2RT \ln c_2/c_1}{(1-\bar{v}\rho)\omega^2(x_2^2 - x_1^2)} \tag{1.12}$$

Thus concentration measurements at two levels in the cell at distances x_1 and x_2 from the centre of rotation could lead to a molecular weight.

The first ultracentrifuges

Just as there were the two theoretical approaches, so were two main types of ultracentrifuge developed. An early low speed (equilibrium) ultracentrifuge was made from a cream separator, but later direct drive machines were built that gave less vibration and by 1933 were running at 18,000 rev/min to give a field of 19,000*g*. The rotor spun on a vertical axis, was cooled by hydrogen and was surrounded by a water bath.

High speed machines were driven by oil turbines. By 1934 fields up to 900,000*g* were being achieved by Svedberg and his associates using small rotors, but these usually exploded after a very few runs. Larger rotors proved to be more accurate and fields in the region of 260,000*g* were more suitable for routine work.

The design of an analytical ultracentrifuge is really dictated by the cell. Sector-shaped cavities are essential for velocity work to eliminate collision reflection from the sides of the cell but not for equilibrium work where they can actually increase the time taken to reach equilibrium. Too short a sedimentation path is inconvenient since back diffusion from the bottom of the cell interferes with observations. On the other hand if the cell has too long a path the windows are too large for strength, hydrostatic pressures are greater, convection is more troublesome and radial dilution effects are greater. In practice cells are now usually arranged to give a sedimentation path of 14 mm or less. Resolving power is directly proportional to the radius of rotation, sedimentation pathlength and the square of the angular velocity. The optical pathlength in the cell must be great enough to provide sensitivity in detecting small concentration of solute. A 12 mm path is common for cells, and this dictates the details of optical design. The optical pathlength now covers a range from 1½ to 30 mm.

Early rotors were circular in order to provide better thermal equilibration. This circular shape is still used today in some modern instruments, but for high speed work oval rotors with two cell holes were found to give the best stress distributions. A good ratio of tensile stress to density is important in the rotor material, but it must be remembered that there is a limit beyond which a cell cannot be made any lighter; hence Svedberg's rotors were made of steel rather than of the aluminium alloys available at the time. However, other designers whose names are mentioned below preferred to develop rotors based on alloys of aluminium.

An atmosphere of low pressure hydrogen was used to remove heat, since vacuum operation was not considered to be satisfactory when heat was being produced from the bearings. White metal plain bearings of small radius to reduce surface velocity were used and the rotor was balanced statically and dynamically to minimize bearing pressures. In later models, a radiation thermocouple sensed the rotor temperature, which was regulated by circulating water round the rotor chamber and by controlling the temperature of the oil. High speeds were measured by a stroboscopic arrangement or by measuring the frequency of alternating current produced in a coil near a magnetized element in the shaft.

The optics used were not entirely the same as are encountered on modern machines. Ultraviolet absorption methods were used and also a black line Toepler–Schlieren system. For equilibrium work, a slit method and Lamm's scale method have been used but are now obsolete; Svedberg and Pedersen (1940) should be consulted for more details on these systems.

Considerable attention was paid to the design of rotor casings, retaining bolts and the concrete pyramid on which the oil turbine driven rotor was supported. It was calculated, for example, that the explosion of rotor VII at 78,500 rev/min would involve forces comparable with the maximal force acting on the projectile of an 8 inch gun.

Another approach to the design of the ultracentrifuge was being developed during the same period as Svedberg's researches. Henriot and Huguenard (1925) obtained high speeds without bearings by using air jets to support and drive a rotor. The development of a successful ultracentrifuge based on the air turbine principle has been associated with the names of Bauer, Beams, Pickels and others. Pickels and Beams (1935) were able to obtain high speeds of rotation with large rotors while maintaining a vacuum in the rotor chamber; the drive was via a piano wire suspension passing through an oil gland—a system very similar to that used in modern ultracentrifuges. Aluminium alloys became commonplace for the making of rotors (Bauer and Pickels, 1937) since it was easy to machine from the manufacturer's stock without extra heat treatment. The metal could flow a little under tension and ease stresses, and, in the event of an explosion, its light weight was a further advantage.

Some early results

Among the vintage results obtained was a molecular weight of 4,930,000 for the haemocyanin from *Helix pomatia* (Svedberg and Chirnoaga, 1928). This was determined with a low speed gear-drive equilibrium ultracentrifuge at 11,000 rev/min and 5,400*g*. Previous estimates of the molecular weight had been in the region of 200,000 and these had been considered enormously high! The gear-drive machine was also used on carboxy-haemoglobin (Svedberg and Fahreas, 1926) and gave a molecular weight of 67,870, thus supporting the osmotic pressure work of Adair which gave an answer of 66,700 for 10 different haemoglobins. Estimates of the molecular weight of haemoglobin based on the iron content had given a minimal answer of 16,700.

The later oil turbine machine was used to clear up problems in temperature control and other technical matters, such as the best rotor design. However, early work was carried out on the digestion of egg albumin by papain (Svedberg and Erikson, 1933, 1934), and many other measurements were to follow. It was soon realized that proteins have discrete molecular weights with the possibility of some aggregates rather than a distribution of molecular weights. An early observation was that protein molecular weights appeared to fall into groups that appeared to be simple multiples of 17,600; this idea was later rejected as more results of greater accuracy became available.

REFERENCES

Adair, G. S. (1925) *Proc. R. Soc., A*, **108,** 626.
Adair, G. S. (1928) *Proc. R. Soc., A*, **120,** 573.
Bauer, J. H. and E. G. Pickels (1937) *J. exp. Med.*, **65,** 565.
Dumansky, A. and coworkers (1913) *Kolloid.-Z.*, **12,** 6.
Henriot, E. and E. Huguenard (1925) *Compt. Rend.*, **180,** 1389.
Millikan, R. A. (1913) *Physik. Z.*, **14,** 796.
Millikan, R. A. (1913) *Phys. Rev.* (2), **1,** 218.
Pickels, E. G. and J. W. Beams (1935) *Science*, **81,** 342.
Pickels, E. G. and J. W. Beams (1935) *Phys. Rev.*, **47,** 336.
Stokes, G. G. (1856) *Trans. Camb. phil. Soc.* Pt II, **9,** 8.
Svedberg, T. and E. Chirnoaga (1928) *J. Am. chem. Soc.*, **50,** 1399.
Svedberg, T. and I.-B. Eriksson (1933) *Biochem. Z.*, **258,** 1.
Svedberg, T. and I.-B. Eriksson (1934) *J. Am. chem. Soc.*, **56,** 409.
Svedberg, T. and R. Fahraeus (1926) *J. Am. chem. Soc.*, **48,** 430.
Svedberg, T. and K. O. Pedersen (1940) *The Ultracentrifuge*, Oxford University Press, Oxford.

CHAPTER 2

The modern ultracentrifuge

The main purpose of this chapter is to give only the briefest outline of the characteristics of modern ultracentrifuges. It is not intended that this should be a catalogue or even an instrument manual since these are best provided by the suppliers of the machines whose addresses are given below. Furthermore, detailed changes are constantly occurring in design and are reported regularly to those whose names appear on their mailing lists.

Without a doubt the Beckman Model E ('Spinco') analytical ultracentrifuge has dominated the laboratories of the world and the author's experience has been mainly with this instrument dating from the early 1950's. It has an advantage in that early models can be upgraded with additional components at any time. This policy has some benefits and disadvantages, but the production of a completely redesigned machine would not fit in with a policy of continuous improvement. In more recent years analytical ultracentrifuges made by other firms have been gaining ground; Bowen (1966) has described some of the features of modern equipment in a general survey of laboratory centrifuges. This chapter cannot be concerned with the niceties of laboratory finance since it is intended for beginners who may not be concerned with such matters. However, it must be mentioned that ultracentrifuges need a regular supply of parts in the form of cells, rotors and reconditioned drive units. Machines made by different companies may be competitively priced initially but the actual maintenance costs may vary considerably; even the locality may have some bearing on these costs.

Although the present-day ultracentrifuge looks vastly different from the pre-war machines mentioned in the first chapter, the main dimensions of cells and rotors were established then and are substantially similar today. Speeds attainable are also similar except where modern titanium rotors are in use. The main changes have been in the drive, speed control, temperature control, optical systems and convenience of operation. Many cells are now available to cover special techniques, and these will be mentioned in later sections of this book. Ultracentrifuges can be classified as analytical or preparative; this is a poor distinction since some preparations are possible in an analytical machine and an optical system is available for at least two makes of preparative machines. However, the apparatus will be considered as belonging to one or other type in the main.

Analytical ultracentrifuges

The Beckman Model E (Beckman Instruments Inc.).

This instrument (Plates I(a), and II(a)) is a familiar sight in laboratories throughout the world and the vast majority of photographs published in scientific journals have originated from it. The rotor is suspended from a flexible wire and only needs balancing to within 0·5 gram. It runs in a vacuum chamber. The drive is from a $1\frac{1}{2}$ HP electric motor geared up $5\frac{1}{3}$ to 1. The direct drive allows a directly coupled tachometer that changes one unit for every 6,400 turns of the rotor so that the speed can be directly measured. From the drive there is a rotating shaft, the speed of which is compared with the speed of a shaft driven by a synchronous motor in a central gearbox which permits the selection of speeds at the rotor. There is also an electronic speed control available for greater precision, particularly at the lower speeds which are now so important for the determination of molecular weights by the equilibrium method.

The rotor temperature is controlled by a rotor temperature indication and control system (RTIC). The rotor has a thermistor in its base and below that a needle touching the surface of mercury to complete an electrical circuit. Variations in the rotor temperature are sensed and made to control an electrically heated coil below the rotor. In order to control at the lower temperatures, say 20°C or less, some refrigeration of the rotor chamber is required, depending on the rotor speed chosen.

The optical systems commonly in use are the Schlieren, Rayleigh interference and ultraviolet absorption systems. Double-beam ultraviolet recording is also available. The next chapter describes these systems in detail.

In view of the fact that this instrument is so well known and that the contents of the rest of the book refer to it constantly (since two of these machines are in the author's laboratories), no further description of it will be given here.

The MSE ultracentrifuge (Measuring and Scientific Equipment Ltd.).

This instrument (Plate II(b)) uses a direct drive unit designed for operation up to 75,000 rev/min. The rotors can be run either in a vacuum or in low pressure hydrogen (Plate III). An infrared sensor placed near the rotor at the level of the cell estimates its temperature, and a feedback circuit enables the rotor temperature to be controlled to better than 0·1°C. The drive unit is a 2 HP triple-phase motor with bearings that are designed to last at least 1,000 hours at full speed. The bearings can be changed independently of the motor. The speed of the motor is controlled by a variable AC generator which dispenses with the need for a gear box. The motor is inside the vacuum chamber so separate seals are not required around the motor spindle.

All the main optical systems are provided and the interference system provides more fringes than usual to assist in their measurement.

The flexibility of the optical system is achieved by use of interchangeable optical parts located in rotating discs and turrets (Plate IV). At the same time, the slit carrier within the rotor chamber is motorized so that the operator can select at will the slits required for the appropriate optical system, even during the course of the experiment. Immediate changeover from the Schlieren method to the interference method or *vice versa*, for example, can be effected without halting the experiment. The two light sources—white light (tungsten lamp) and monochromatic light (mercury lamp with monochromatic filters)—are selected merely by turning a mirror. The interference and ultraviolet filters are selected simply by rotating a disc, as are both of the objectives (Schlieren lenses) and the inserts (phaseplate or horizontal knife edge).

All the optical systems, including the cylindrical lens system, are fully corrected for both visible and ultraviolet light, so that refocusing is completely avoided for all methods. Indeed, all optical parts are precisely prefocused for maximum sharpness of diagram, whichever optical system is used. The design of the optical system permits one single optical path to be used for all known optical techniques. Changeover from one method to the other is consequently extremely simple and is achieved without halting the experiment.

A variant of the Schlieren system (Toepler–Schlieren) is being developed to allow concentrations as low as 0·005 per cent to be measured by visible light. Unlike the very sensitive absorption method this does not depend on the selection of an absorption maximum for the component. This has a further advantage in that it provides Gaussian curves directly, whereas absorption plots determined densitometrically on photographic material have to be differentiated to obtain this information. An ultraviolet scanning system is also available for the instrument (Plate V(a)).

Although the author has not personally used the MSE analytical ultracentrifuge he has heard good reports of its smooth running characteristics even at low speeds. The makers claim that this makes it possible to obtain diffusion data using a synthetic boundary cell with a high degree of precision. The arrangement of optical system and rotor chamber appears to be at a convenient working height for the operator.

The Christ ultracentrifuge (*Martin Christ, W. Germany*)

The author has not had direct information on this instrument from its users, but Christ preparative machines have long enjoyed a reputation for high quality (Plate V(b)). The following notes were taken from the firm's literature.

The apparatus is driven by an electric motor. The impetus is transmitted by a flexible connecting link to the rotor axle, which runs without vibration in a specially developed oil pressure chamber. All rotors are easily set up without any screwing to the rotor axle. The oil pressure chamber is supplied by a hydraulic unit with motor-driven pump, pressure reducing valve, cooler and filter. This unit also supplies the hydraulic cylinder for the raising and lowering of the rotor chamber. This is protected against a possible breakdown of the rotor by multilayer armour. In addition, the chamber incorporates the heating and cooling units necessary for the temperature regulation of the rotor. Directly connected to the rotor chamber is the oil diffusion pump which, together with the two-stage rotary valve pump, produces a high vacuum of about 10^{-3} torr. A refrigerated unit prevents oil residue diffusion and the coating of the optical parts inside the rotor chamber.

A noteworthy innovation is the arrangement of the thermometer in the middle of the rotor in a centrifuge with an upright rotor. This was made possible by the successful development of a reliable transmission system, which records the measurements of the moving rotor. Transmission results from the action of two concentric induction coils fitted between the rotor and the motor, the inner one on the rotary secondary winding of the rotor axle and the outer one on the fixed primary winding of the casing, which is connected as a branch of the thermometer bridge. During the analysis the temperature is kept constant at 0·1°C, which is achieved by a thermoelectric cooling system working by means of proportionally controlled Peltier batteries. The temperature range of $-5°$ to $+25$°C includes all possible preparative and analytical work and the rotor temperature remains constant at about 0·05°C. The revolution count is adjustable continuously between 1,500 and 60,000 rev/min. The apparatus is fitted with Schlieren, interference and ultraviolet optics. The interference method allows the determination of strongly diluted solutions. The interferogram also appears as Gaussian distribution curves. The Schlieren method can be easily reversed to interferometry by moving a lever. The ultraviolet optics, with which the apparatus can be fitted, are independent of all other optical combinations. The ultraviolet absorbance can be automatically registered with a recorder or photographically with 10 or 26 exposures. The gradients can be registered by a recorder as Gaussian distribution curves during sedimentation. The registering instrument is automatically combined with the photographic apparatus, in order to make use of both possibilities at the same time. In addition to the ground glass plate the course of the sedimentation can be followed with a monitor. For the photographic exposures, 2×10 inch plates are used which make six exposures possible, and the series of exposures can be automatically regulated.

In addition, the ultracentrifuge can be equipped with a diffusion unit which works with a large number of cells.

The Metrimpex ultracentrifuge (Metrimpex, Hungary)

The rotor is driven by an air turbine. The advantage is claimed to be that there is little to wear out except the shaft and bearings, which can be replaced by the user. Speed is read out from a compact frequency generator on the rotor shaft. The temperature control is based on a thermistor in the rotor and the rotor spins in a vacuum. Balancing has to be within 0·5 gram and the three main optical systems are available for observation and measurement.

Preparative ultracentrifuges

In preparative work a compromise has to be made in the choice of rotor. Ideal sedimentation occurs in sector-shaped cavities designed to minimize convectional processes. More will be said on this subject in Chapter 9. For rapid harvesting, angle rotors are best since they exploit a short path-length to the tube wall; a heavy layer forms here which causes convection currents. In this way material can be rapidly deposited as a pellet, although the convection can mitigate against the resolution of difficult mixtures of closely sedimenting material. The swinging bucket rotor provides more ideal conditions but the tube is not the ideal sector shape and there can be swirling effects as the buckets swing at the start and end of the run. However, such rotors are very popular in conjunction with density gradient techniques. The Anderson rotors employ true sectorial cavities which can be loaded or unloaded with the rotor actually spinning. The Anderson rotors are particularly valuable for the fractionation of large

(a)

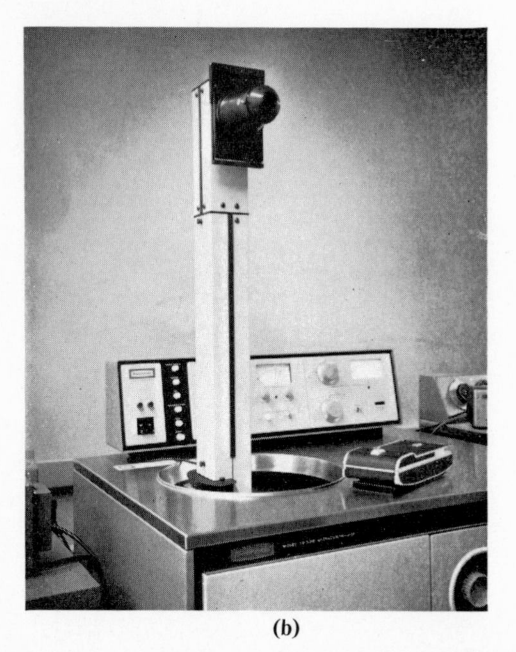

(b)

Plate I(a). The Beckman analytical ultracentrifuge

Plate I(b). The Beckman Schlieren optical attachment for use with preparative ultracentrifuge (Beckman Instruments Inc.)

[*Facing page* 12

(a)

(b)

Plate II(a). A later version of the Beckman ultracentrifuge shown in Plate I(a). The instrument has been raised to accommodate a monochromator and high intensity light source. The instrument has been equipped with a photoelectric scanner and a multiplex unit that permits particular cells in a multicell rotor for scanning. To the right of the instrument stands a data interface accessory for computerization of the results (Beckman Instruments Inc.)

Plate II(b). General appearance of the MSE analytical ultracentrifuge

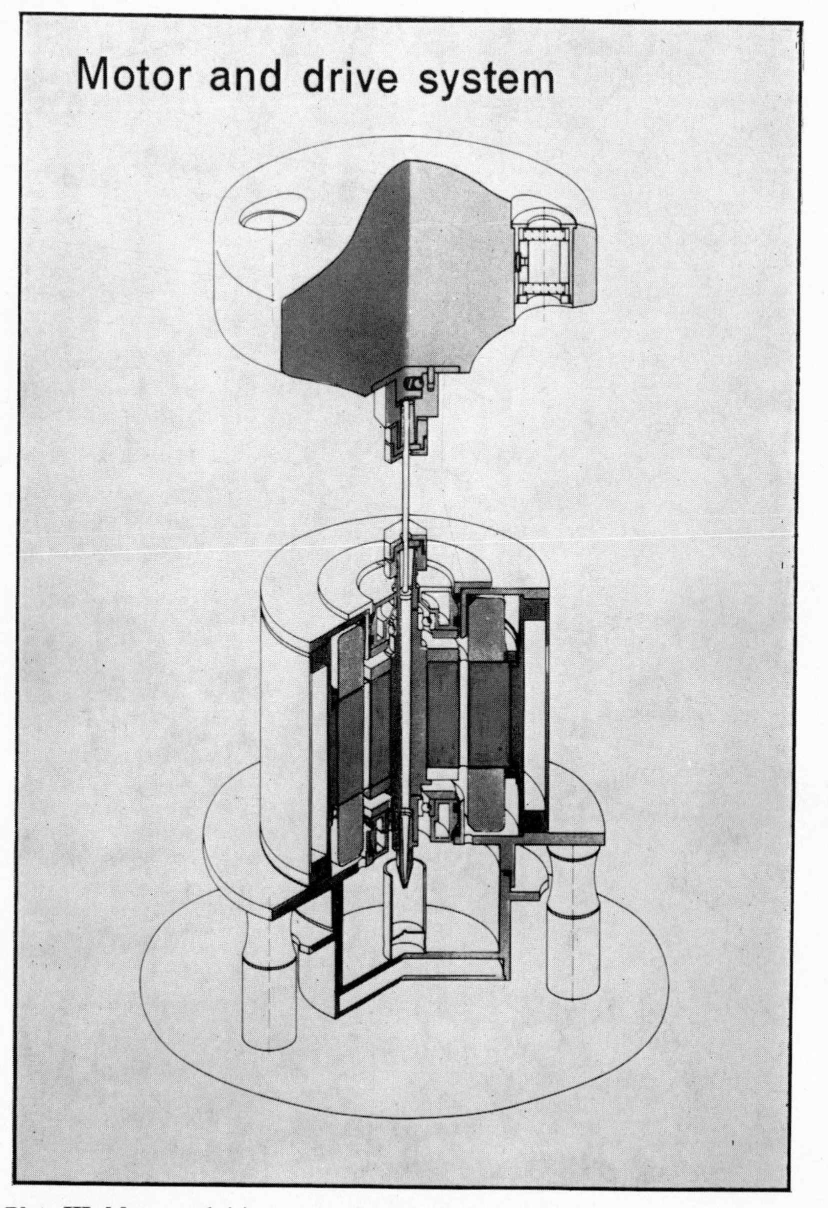

Plate III. Motor and drive system for the MSE analytical ultracentrifuge. Both rotor and drive motor are in the chamber, where they may be run in a vacuum or in low-pressure hydrogen

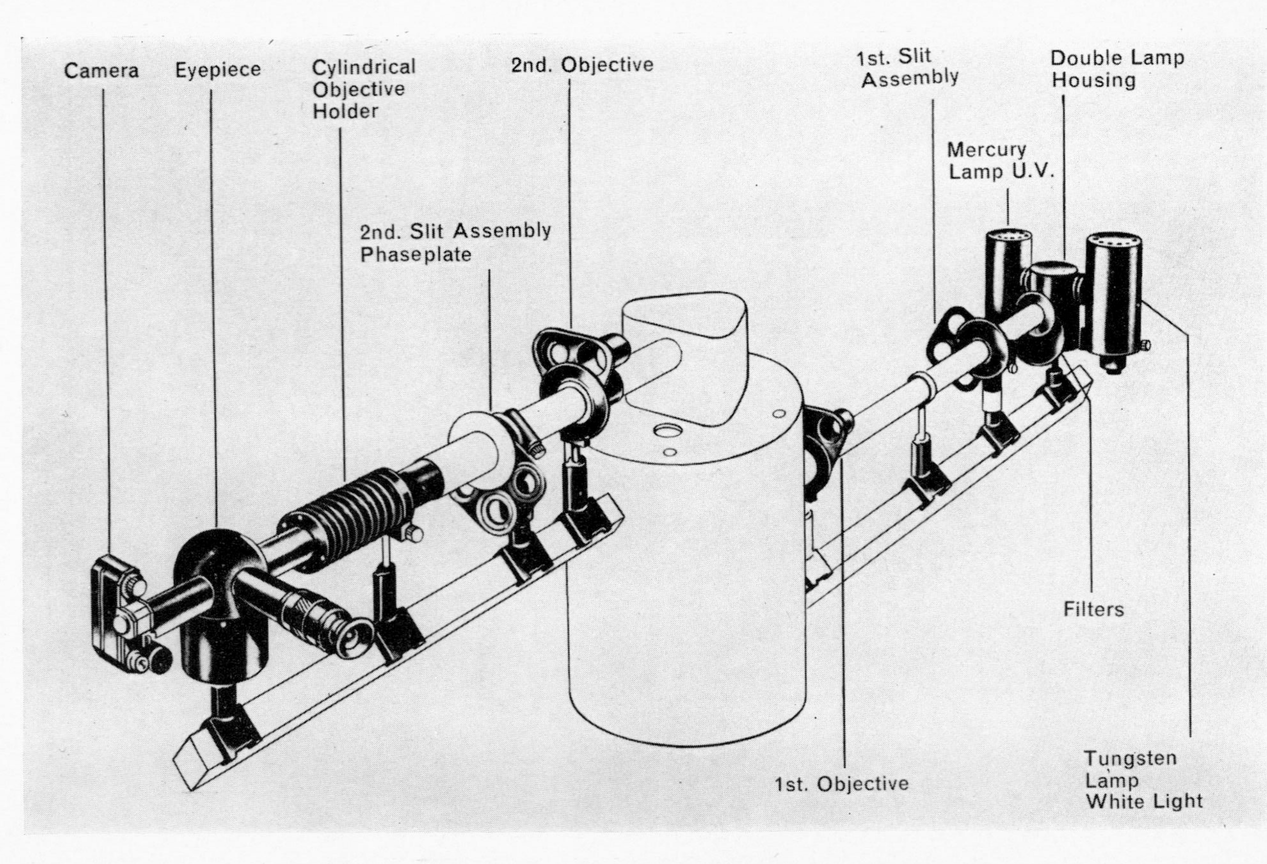

Plate IV. Optical arrangements for the MSE analytical ultracentrifuge

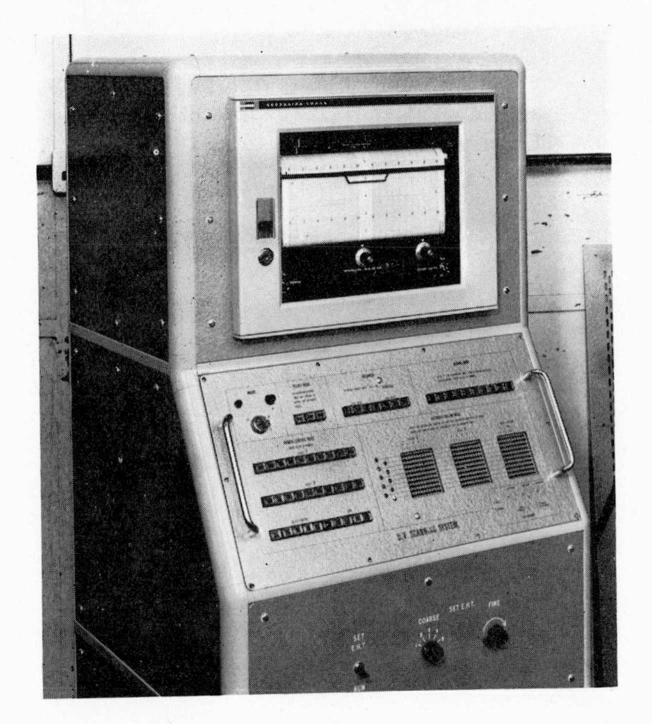

(a)

(b)

Plate V(a). Ultraviolet scanning system for use with the MSE analytical ultracentrifuge

Plate V(b). The Martin Christ analytical ultracentrifuge

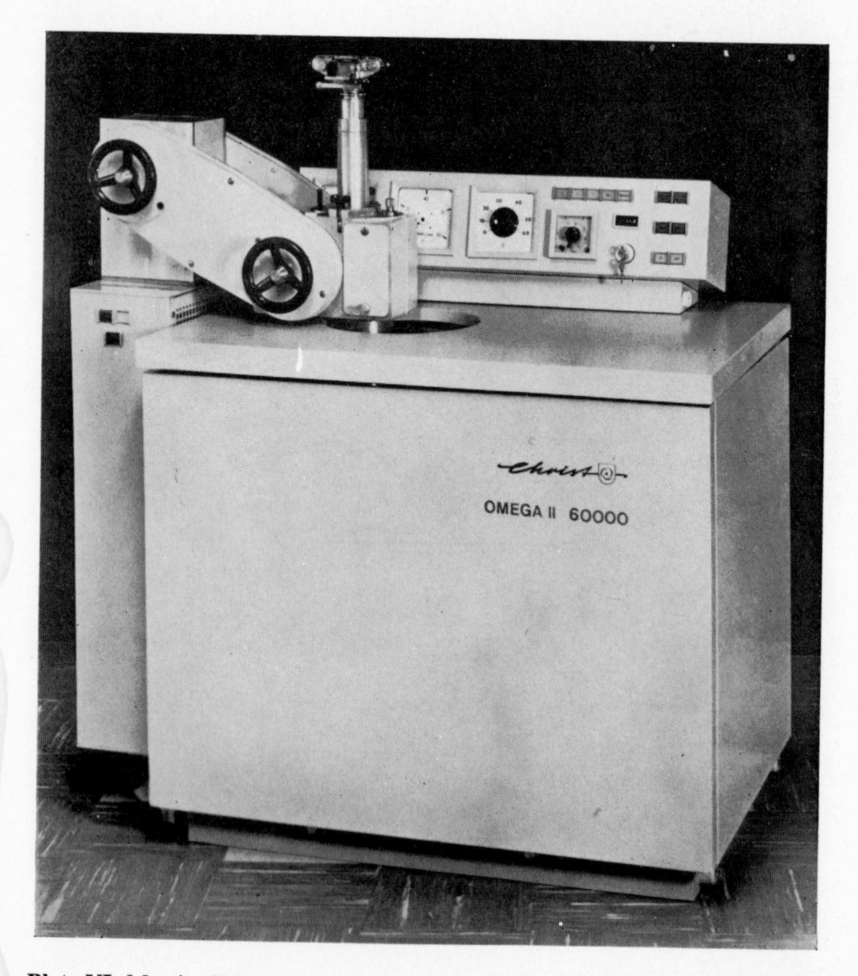

Plate VI. Martin Christ preparative ultracentrifuge to which an optical system (Schlieren and interference) has been fitted so that analytical data may be obtained

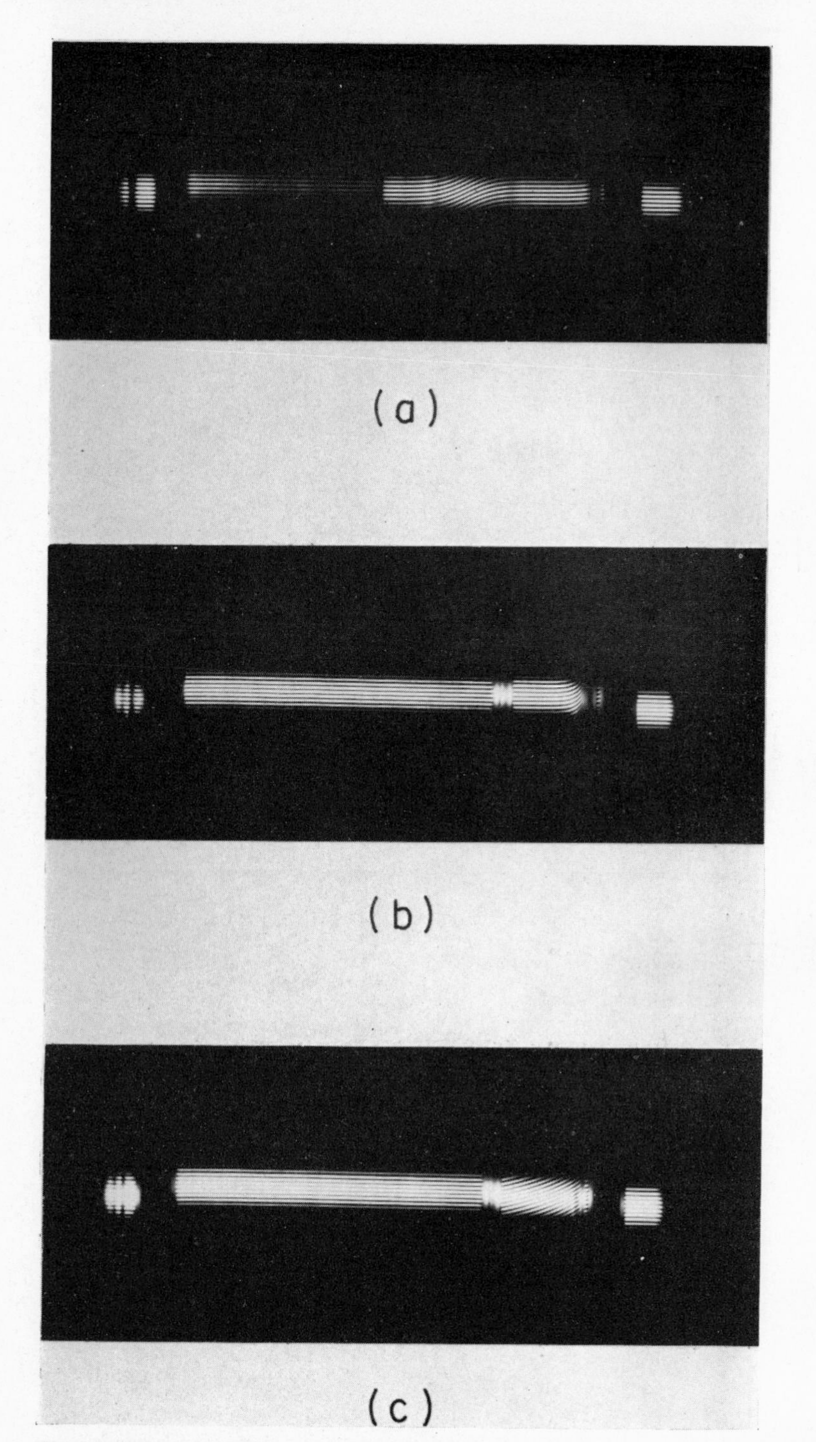

Plate VII. Interference photographs taken with the Beckman Model E ultracentrifuge to show the appearance of fringes depicted in Figure 6.2. (a) is a late picture from a synthetic boundary run, (b) is a meniscus depletion run, and (c) is a low speed equilibrium run

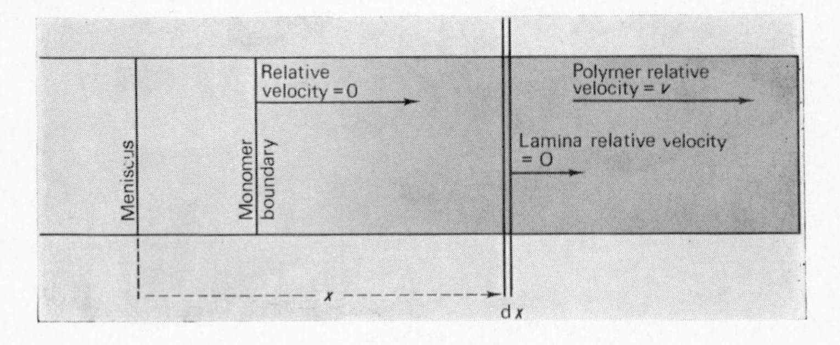

(a)

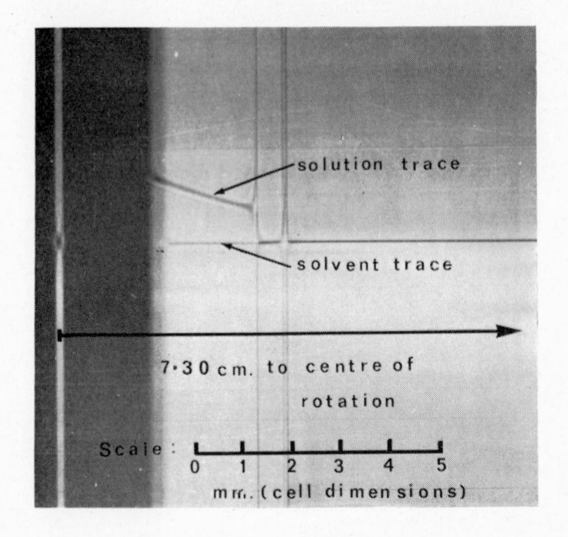

(b)

Plate VIII(a). This drawing defines the distances and relative velocities mentioned in the text

Plate VIII(b). Schlieren trace from an equilibrium experiment on a serum protein, using a double sector cell. Speed = 8,095 rev/min. Assume $\rho = 1{\cdot}004$, $\bar{v} = 0{\cdot}73$, $T = 7°C$, $R = 8{\cdot}3 \times 10^7$ cgs units for calculation purposes. (Because of distortions involved in reproduction, only a limited degree of accuracy can be achieved)

volumes of material in density gradients under near-ideal conditions. Most of the major manufacturers vie with each other to be the first to supply the highest g values with the largest capacities or the longest path-lengths to give high resolution in density gradient work. Titanium rotors are offered to provide the highest gravitational fields. In choosing a preparative ultracentrifuge one has to consider the price, the range of rotors offered, the temperature control and the freedom from vibration. Temperature control has to be very precise and reproducible, particularly when working near the freezing point of the solutions.

Rotors are usually mounted on a flexible suspension and can wobble at low speeds. Beckman ultracentrifuges provide a mechanical arrangement to support the rotor at the beginning and end of the run to prevent the wobble occurring. Other makers claim that the problem is obviated in other ways. In the MSE centrifuge, wobble is said to be obviated by care in the positioning of the oil seals and in the correct sizing of the drive shaft and other rotating parts. The Christ machine uses a special hydraulic damping device and an electrical relay which precludes the supply of full power to the drive motor at low speeds.

For the most careful work preparative machines employ a diffusion pump in addition to a mechanical pump in order to minimize friction with residual air.

The modern analytical ultracentrifuge is becoming more and more elaborate and expensive to buy and maintain. The development of the double-beam ultraviolet scanning optical system alone (Chapter 3), although highly desirable, is a very expensive addition. On the other hand, in a typical department of biochemistry many of the runs carried out are merely to check on the size homogeneity of a preparation using Schlieren optics. The information obtained on the sedimentation coefficients, and the relative abundance of the components present can then serve as a guide to the intelligent planning of a preparative run. The Martin Christ preparative ultracentrifuge can now be provided with an optical system (Plate VI) to provide such information at a much lower cost than would be involved in the purchase of a conventional analytical ultracentrifuge. The Beckman Instrument Company have also developed such an attachment, and this would appear to be a most desirable development for those departments that have neither the finance nor the workload for the traditional analytical machine. Alternatively, such a device liberates the more expensive machine for the more biophysical type of work for which it is specially equipped.

It is possible to use preparative rotors in an analytical ultracentrifuge. However, very few departments are likely to tie up their valuable analytical instrument for this purpose. Preparative work is best kept for preparative ultracentrifuges.

ULTRACENTRIFUGE MANUFACTURERS

Beckman Instruments (Spinco Division) Inc., Palo Alto, California, U.S.A.

Heinz Janetzki, K.-G., 7123 Engelsdorf (Bez. Leipzig), Leipziger Strasse 106–112, East Germany.

Hitachi Ltd., Tokyo, Japan.

International Equipment Company, 300 2nd Avenue, Needham Heights, Massachusetts, U.S.A.

Ivan Sorvall Inc., 100 Pearl Street, Norwalk, Connecticut, U.S.A.

Lourdes Instrument Corp., 656 Montauk Avenue, Brooklyn, New York, U.S.A.

Martin Christ, 3360 Osterode Am Harz, Postfach 1220, W. Germany.

Measuring and Scientific Equipment Ltd., Manor Royal Crawley, Surrey, England.

Metrimpex, Hungarian Trading Company for Instruments, Budapest, 62, Hungary.

Note: Analytical ultracentrifuges are not manufactured by all the companies listed but all can supply preparative instruments.

REFERENCE

Bowen, T. J. (1966) '*Instrumentation in Biochemistry*', Biochemical Society Symposium No. 26, Academic Press, London and New York.

CHAPTER 3

Optical systems

A whole chapter must be devoted to the optical systems used on a modern ultracentrifuge. In general one expects to find three main systems. Ultraviolet absorption was used as a method of studying the migration of a solute boundary from the earliest days; it is available in single or double beam forms and with photographic or chart recorder read-out. Refraction methods are available also either as a Philpot–Svensson Schlieren system or as a Rayleigh interferometer. At present, all three systems have their uses, although the latest double beam ultraviolet systems with chart read-out may eventually be the only system required.

During sedimentation the concentration of a solute increases towards the bottom of the cell. A direct graph of the concentration against x is often required and is provided by the ultraviolet absorption system or by Rayleigh interferometry. A derivative plot of dc/dx against x is, however, useful for distinguishing closely-spaced components and is the form required in some mathematical equations. Such a derivative plot is obtained from the Schlieren system and can also be obtained using electronic circuitry from the ultraviolet double beam recording absorption system.

The ultraviolet absorption system is the most sensitive and has detected concentrations as low as 0·001 per cent of DNA (Shooter and Butler, 1956; Aten and Cohen, 1965). A modern system utilizes a monochromator so that the optimal wavelength can be selected for a particular component of a mixture. The ability to detect such low concentrations is valuable when one is trying to work with solutes whose parameters (e.g. sedimentation coefficient) cannot be easily extrapolated back to zero concentration. The Rayleigh system is quite sensitive and is of particular value in sedimentation equilibrium studies since it employs two channels, one for solvent and one for solute; this enables the effect of salt redistribution and window distortion to be partially cancelled, but a dummy duplicate run is still required to estimate the contribution made by cell distortion. The Philpot–Svensson (cylindrical lens) modification of the Schlieren system is the most frequently used and is the only system based on refractivity that can be adjusted to give the best picture while a run is in progress. It is the least sensitive of the three systems but is used for most of the pictures seen in the scientific journals.

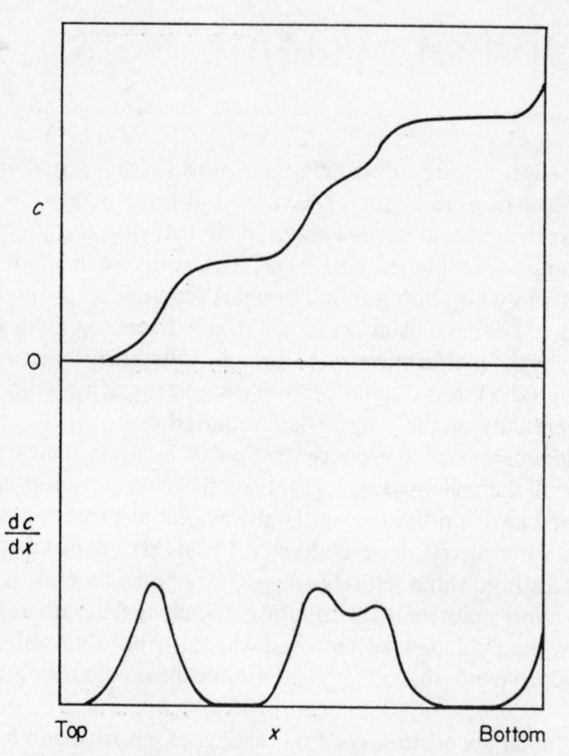

Figure 3.1. Two methods of showing sedimentation data. The direct plot of c against x (above) is given by ultraviolet absorption or by Rayleigh interferometry. The lower plot of $\mathrm{d}c/\mathrm{d}x$ against x arises from Schlieren optics. Back diffusion from the bottom of the cell is usually observed in practice, as indicated in the diagram

Interferometric methods measure differences in refractive index between a solvent and a solute channel. Schlieren systems measure the gradient of refractive index. Thus, in Figure 3.1, the refractive index difference between solute and solvent, $n_s - n_0$ is actually given by Rayleigh interferometry instead of concentration, c, in the upper curve. Schlieren optics plot $\mathrm{d}n/\mathrm{d}x$ instead of $\mathrm{d}c/\mathrm{d}x$ as shown in the lower curve. A direct replacement of refraction data for concentration must bring in a proportionality constant, k, known as the specific increment. Thus:

$$n_s - n_0 = kc \qquad (3.1)$$

The equation is only valid when concentration is expressed as weight per volume of solution (usually g/100 ml). This linear relationship is usually reasonably valid for macromolecules such as proteins (Lamm, 1937).

The absorption system

The laws of light absorption are those already familiar in the field of spectroscopy. If I_0 is the intensity of the incident light and I the intensity after transmission through a cell of optical path, d, containing a concentration of solute, c, then, if Beer's law is valid:

$$\text{optical density} = \log_{10}\frac{I_0}{I} = \varepsilon cd \qquad (3.2)$$

where ε is the molar extinction coefficient. If a single beam system is used then a suitable calibration of the photographic plate in terms of blackening against concentration has to be prepared. Such a single beam system suffers when oil mist is deposited on the optical surfaces from variations in the intensity of the light source and changes in the photographic technique.

Nowadays the double system is greatly to be preferred, although it is very costly. The Beckman apparatus involves a monochromator below the ultracentrifuge. A photoelectric scanner using the split beam principle enables sample and its solvent to be compared in the two channels of a double sector cell. A double sector cell (as shown in Figures 3.5 and 3.9) has two channels for solvent and solution respectively so that in this instance it behaves in an analogous manner to the solvent and solution cells used in conventional spectroscopy. Calibration steps are recorded before each scan and an optical density range of 0–1 or 0–2 can be covered. Three scanning speeds are provided and the absorption due to solvent can be automatically subtracted from the solution trace. A derivative plot is also available. Such a system enables sedimentation equilibrium estimates of molecular weight to be made at very low concentrations with increased accuracy, since only extrapolated values are ideal and give the correct answer.

The most exciting possibility for the future is the extension of the electronic circuitry to give a read-out suitable for computer technology. Trautman (1964) has referred to speed, temperature and other parameters as being relegated to the computer data input. A molecular weight result should be given directly using all the data necessary automatically. In England, the Birmingham group have been particularly active in the application of computer technology to the ultracentrifuge. Spragg and his coworkers (1963) have built their own version of a double-beam ultraviolet system. The ultraviolet system may make its most important gain in accuracy over the Rayleigh system when the chart record is eliminated in favour of a digital presentation.

The Schlieren system

The first thing that one should say to English-speaking students about a Schlieren system is that Schlieren was not a scientist. In fact, the word is German for streaks, such as one sees in imperfect glass. The Schlieren method has many variants. To understand its working let us start by considering a simple example. Take an illuminated slit and use a lens to bring the light to an image of the slit. Now take a second lens, a camera lens, beyond the image of the slit and take a photograph of the first lens. This photograph would be a uniformly illuminated picture of the first lens. Now start again but assume that the first lens had a fault in it so that it did not image all the light from the slit correctly but allowed some stray rays to fall below the image formed of the slit. If the camera is now used to photograph the faulty lens, a uniformly illuminated picture of that lens is obtained providing all the light, including the stray rays, got through the camera lens. The next stage in the argument is to intercept the stray rays below the true position of the slit image before taking the photograph. The photograph now shows a dark patch, due to the intercepted rays, showing where the fault lay in the lens. This is a well-known and highly sensitive method for the testing of lenses. The Schlieren optical system can be argued as an extension of the above.

At one time a 'scanning' Schlieren optical system was found on earlier forms of electrophoresis apparatus. The cylindrical lens Schlieren now found on analytical ultracentrifuges is a more convenient development. However, it may be easier to describe the scanning method first, since it is more easily represented in two dimensions, and then extend the argument to the cylindrical lens modification.

Figure 3.2 illustrates the essentials of the scanning Schlieren system. We have an illuminated slit, a lens which we now call the Schlieren lens, and an image of the slit formed at X. The Schlieren lens has no faults, but we have placed near it a cell in which there is an optical 'fault' in that the cell contains solvent at the top forming a boundary with solution below. B marks the plane of maximum concentration (i.e. refractive index) gradient while A and C represent two other planes where the concentration gradient is equal but, of course, less than at B. Rays suffer maximal

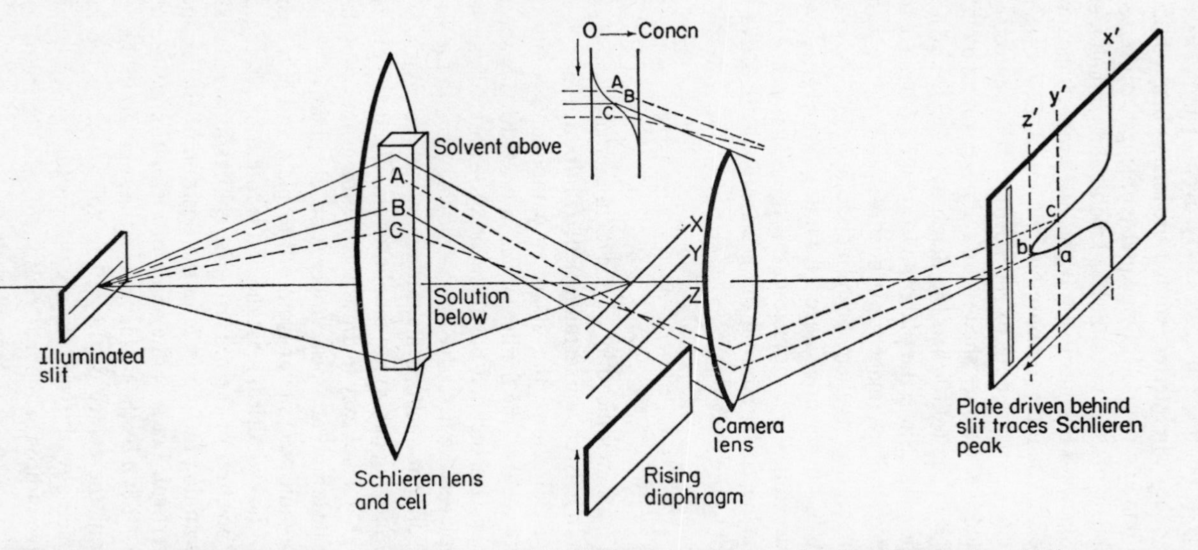

Figure 3.2. A scanning Schlieren system (not to scale). Schlieren lens images the illuminated slit at X if no rays are deflected by a boundary in the cell. The camera lens focuses the cell on the plate. The inset diagram represents the variation of concentration, across the boundary: B has the maximum value of dc/dx and will deflect the ray most, but pairs of planes can be found at A and C that will deflect rays to the same focus

deflection at B to give the image at Z while rays from B and C meet at Y where, incidentally, they can interfere so that the images below X actually show interference banding due to the unequal pathlengths traversed. Let us move a horizontal diaphragm up in the focal plane while, simultaneously, moving the plate sideways behind a slit. If the diaphragm is below *Z* no light is intercepted. When the diaphragm reaches Z a dark shadow appears on the plate corresponding to the position B in the cell. When the diaphragm reaches Y the dark shadow on the plate corresponds to the region from A to B in the cell. When the diaphragm reaches X no light at all can reach the plate. Since the plate moved continuously behind the slit we have produced a black image (on a positive print) in the form of a peak corresponding to the gradient or refractive index in the cell. The scanning method is inconvenient because the whole picture is not available until the plate is developed. The adjustment of the height of the peak on the plate can be achieved by altering the gear ratio in the drive to the diaphragm relative to the plate drive.

Cylindrical lens Schlieren

The cylindrical lens modification of the Schlieren optical system is known as the Philpot–Svensson arrangement (Philpot, 1938; Svensson 1939), and Figure 3.3 shows the essential components omitting the illuminated slit to the left of the Schlieren lens in order to save space. The cylindrical lens scans the inclined bar from side to side and can be considered to have replaced the necessity for the moving plate in the scanning arrangement. By using an inclined diaphragm to cut the rays x, y and z the necessity for a moving horizontal diaphragm is obviated. One can see, intuitively, although it is not easy to draw in two dimensions, that the same result will be produced in that a peak appears on the plate.

Let F be the magnification factor from cell to plate and C the enlargement factor due to the cylindrical lens; the first is found by photographing a graticule in the position of the cell with the rulings horizontal and the second by photographing the graticule in the position of the inclined bar with the rulings vertical. Let θ be the angle of inclination of the inclined bar to the vertical.* Let a be the pathlength of the solution in the cell and b the distance from the centre of the cell to the plane of the inclined diaphragm; b is actually the sum of all the distances divided by the respective refractive indices of the materials involved. Then:

$$n_s - n_0 = \frac{\text{Area of peak}}{F^2 Cab \tan\theta} \tag{3.3}$$

* On the Beckman Model E the angle of inclination of the phase plate is $90-\theta$.

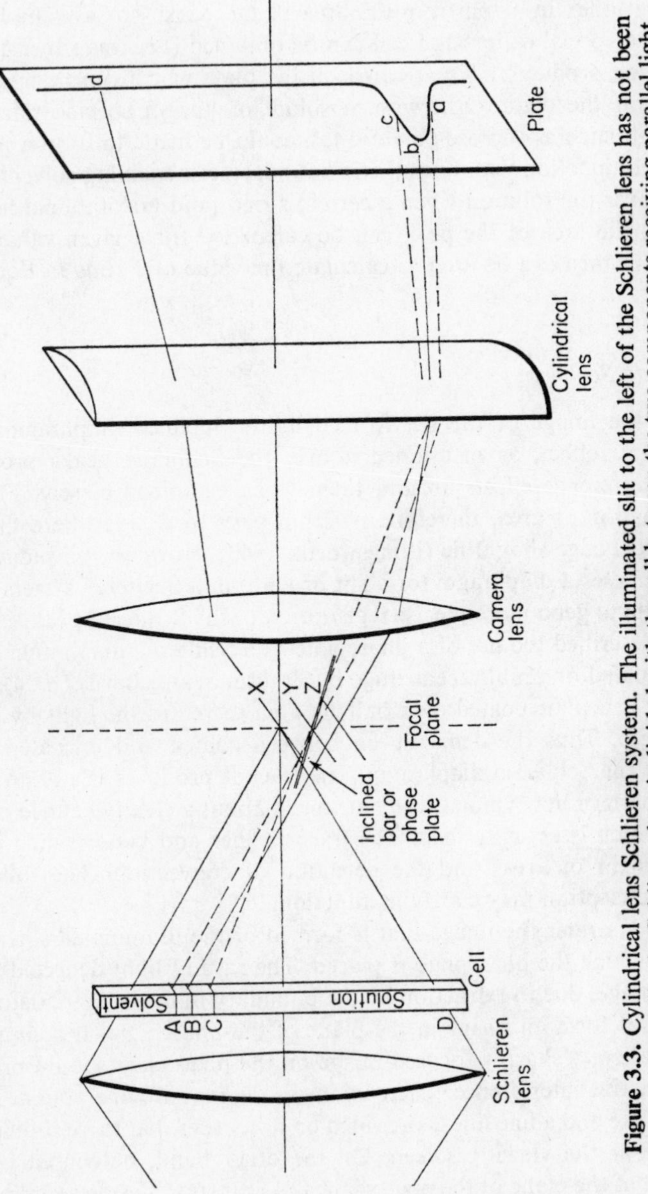

Figure 3.3. Cylindrical lens Schlieren system. The illuminated slit to the left of the Schlieren lens has not been shown. Normally the Schlieren lens is divided with the cell between the two components receiving parallel light. See text for explanation

Thus, knowing the optical constants one can determine the change of refractive index in a cell from the area of the peak. To save the labour involved a special calibration cell can be obtained (Beckman Instruments Inc.) which produces a square area on the plate which can be measured directly off the plate. Otherwise, a solute of known specific refraction increment, such as sucrose ($k = 0{\cdot}143$), could be made to form a peak if used in conjunction with a synthetic boundary cell enabling solvent to be layered over the solution when a certain speed (and gravitational field) is attained; the area of the peak can be calculated for a given value of θ, and this in turn can be used to calculate the value of F^2Cab in Equation (3.3).

Phaseplate Schlieren

Since the image of the slit formed at the inclined diaphragm is an interference effect, as mentioned above, the Schlieren peaks produced show interference effects around them when examined closely. For the determination of area, therefore, it is not easy to decide where the true geometrical edge should lie (Longsworth, 1943). However, nowadays one no longer uses a diaphragm (or a slit or wire) in a Schlieren system but a phaseplate to get a more precise registration of the Schlieren peak. In 1950 Wolter described the use of a phaseplate as a Schlieren diaphragm and it was evaluated on an ultracentrifuge by Trautman and Burns (1954). Such a device is a plate coated over half its area to retard the light by half a wavelength. Thus the demarcation between coated and uncoated areas serves as the Schlieren diaphragm. The trace it produces is still an interference pattern but symmetrically arranged about a clearly defined central region which is actually read. The trace is finer and better suited to the measurement of areas and the detection of convection. The following simple description may clarify its function.

Let us consider the image that is formed of the illuminated slit, at the focus of which the phaseplate is placed. The rays of light deflected below the slit image, due to refraction by the boundary in the cell, are defocused and do not form an image in the plane of the phaseplate. If a simplified view is taken, a sharply-focused image on the phaseplate should produce no destructive interference. There is, therefore, no base line obtained with a phaseplate and a fine line is provided on it in order that there should be a base line at the viewing screen. On the other hand, defocused light is divergent at the plane of the phaseplate and, if it straddles the demarcation region on the phaseplate, will be partly retarded by half wavelength and partly unretarded so that destructive interference can occur before the rays come to a focus on the viewer. Hence the peaks are produced by the phaseplate but the baseline is produced by the fine black line ruled on it. Incidentally, some phaseplates have been found to be defective and unsuitable for use with the Archibald 'approach-to-equilibrium' method for the determination of molecular weight.

Optical alignment

Detailed procedures for the alignment of Schlieren (and Rayleigh) optical systems have been described by Lee Gropper (1964). The Archibald technique demands the precision alignment of the optical system. A tilted Schlieren base line can arise when the slit at the light source is not properly focused at the phaseplate. Skewed peaks, from systems normally known to give symmetrical peaks, can arise if the camera does not focus the cell correctly on the photographic plate. In the above, only one Schlieren lens has been shown but it is common practice to have a lens each side of the cell with parallel light passing through the cell.

Further effects not due to misalignment of the optics are as follows: a raised baseline, caused by a density gradient due to compression of the solvent; distortion, caused by a sloping baseline; a saucer-shaped baseline, arising from redistribution of salts present in the solvent.

The Rayleigh interferometer

Rayleigh (1896) devised an interferometer for the measurement of the refractive indices of the recently discovered gases argon and helium. Light from a slit was rendered parallel by means of a collimating lens and then passed through two slits in front of a sample cell and a reference cell. A further lens focused the light from the two cells with the production of an interference pattern. The original form of the Rayleigh refractometer was not suitable for relating refractive index to the height in the cell as required for use in an ultracentrifuge. Philpot and Cook (1948) showed that the introduction of a cylindrical lens to focus the cell contents in the final image made it possible to obtain a set of fringes each of which gave a curve relating refractive index to the height in the cell. Figure 3.4 shows a suitable arrangement.

Obviously, most of the components of such a system are in common with those of the cylindrical lens Schlieren system. In practice, as regards the Beckman ultracentrifuge, a quick changeover from Schlieren to Rayleigh systems can be made as follows: rotate the light source slit through 90° (or use a push–pull slit), insert a filter in the light path, narrow the light source slit, change the slit above the cell to a double slit, set the phaseplate to 90° and remove the ground glass viewing screen. If an offset slit is in place above the cell, this will serve for Schlieren or Rayleigh work so the change from one system to the other can be made without

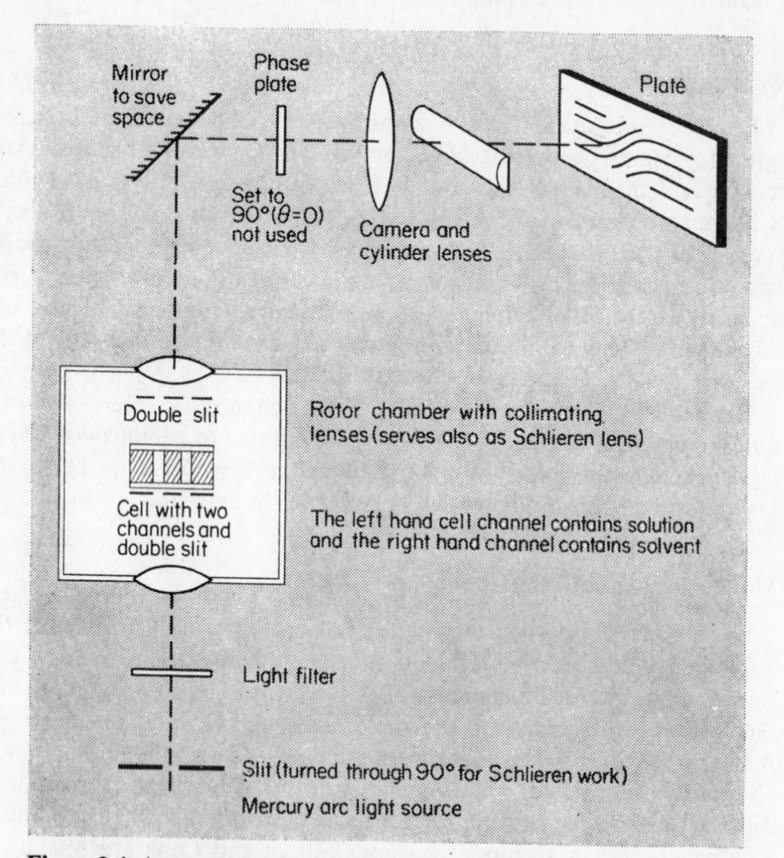

Figure 3.4. Arrangement for producing Rayleigh fringes but using most of the components of the Schlieren optical system (Beckman)

opening the rotor chamber. The offset slit provides one slit on a radius of rotation so that dimensions in the solution channel are dimensionally correct when measured at the plate; this means that the other slit in the pair is not on the radius and therefore not quite at the same reference level as the solution slit. In most work done using the equilibrium method the slight lack of registration between the concentrations of buffer salts in

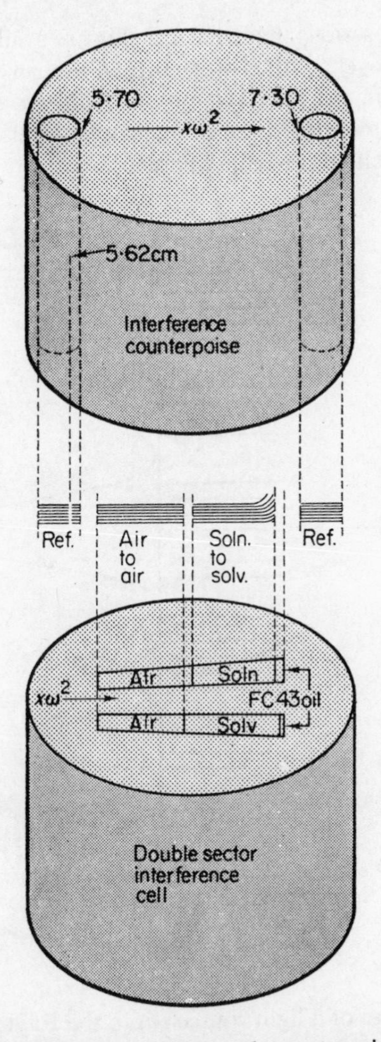

Figure 3.5. As the cell and counterpoise in turn enter the light beam they produce an image on the plate as shown. The counterpoise provides known distances from the centre of rotation to enable, by interpolation, distances in the cell to be determined. Since verticals are not sharply imaged unless there is a point source of light a reference wire (5·62 cm) gives a more precise location of distance than the dimensions of the edges of the holes in the counterpoise (Beckman)

each channel is not serious. Symmetrical slits are available but this means that neither will lie over a radius of the rotor although each will correspond to identical levels in the cell. Symmetrical slits are therefore preferable when it is of paramount importance that similar levels in both channels of the double sector cell are to be compared.

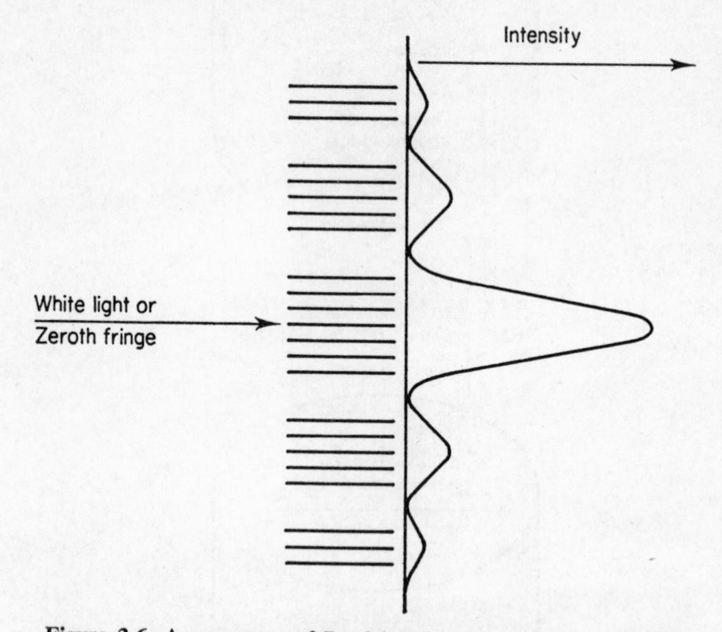

Figure 3.6. Appearance of Rayleigh fringes as produced at the plate. Although the eye can see the weak side fringes, they do not appear on the plate when the exposure is correct for the most intense fringes. If monochromatic light is not used only the zeroth fringe remains sharp

A feature of the use of a light source slit in the Rayleigh method is that verticals are not sharply rendered at the viewer. A point source would give sharper verticals but with a fall in light intensity. The picture obtained (Figure 3.5) represents the centripetal edge of the cell on the left and the centrifugal edge or cell bottom on the right.

The Rayleigh fringes observed are normally about seven in number. Actually there are many more, as shown in Figure 3.6, but when the centre pattern of fringes is correctly exposed the weaker sets of fringes are not shown on the photographic plate. Only one fringe in the centre of the pattern has a position that is independent of wavelength, so this is known as the zeroth or white light fringe. The use of monochromatic light ensures that the other fringes remain sharp. Svensson (1951) has discussed various ways of increasing the number of intense interference fringes observed; one method is to use a raster rather than a single light source

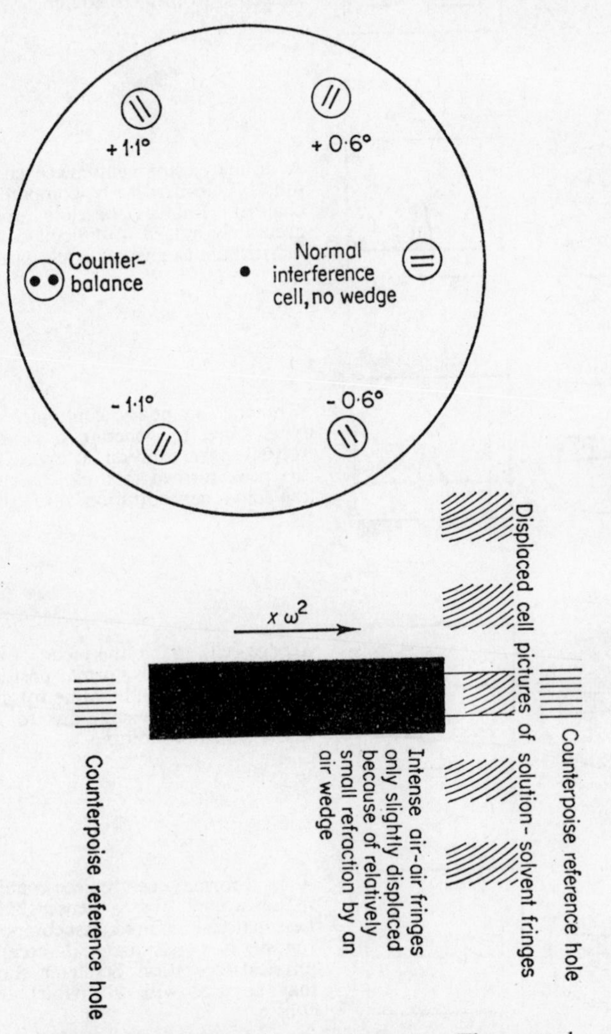

Figure 3.7. Operation of a multicell rotor. The normal non-wedge cell produces the middle picture, but is not filled to such a great depth as the wedge cells in order that the fringes shall not be overlapped by the intense air–air fringes from all five cells (Beckman)

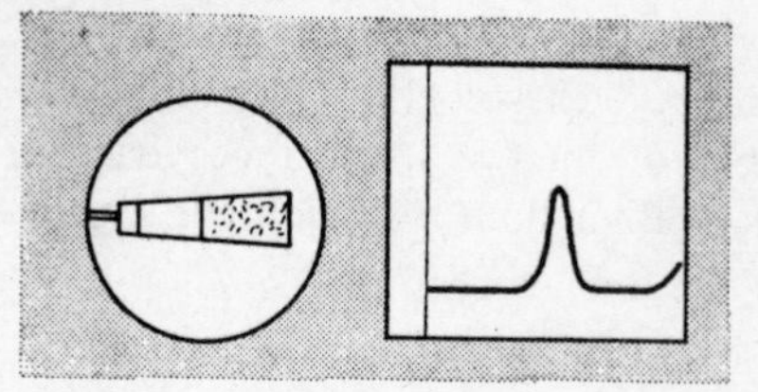

Single sector cell. The workhorse for most Schlieren runs. Usually has a 4° 12 mm centrepiece but other cells can have 2° angle and thickness from 1½ to 30 mm

A double sector centrepiece enables solvent–salts system used to be compared in a second channel. Enables baseline position to be accurately judged in difficult cases where salts redistribute to give a curved baseline

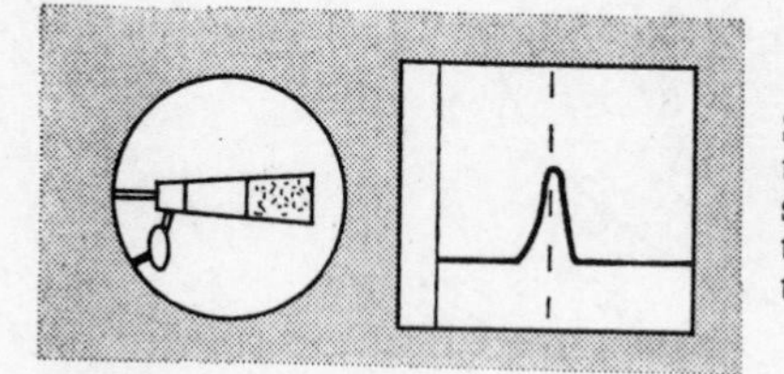

Synthetic boundary centrepiece of capillary type. Used to generate a peak by layering solvent over solution at speed. The area of the peak formed then can be correlated with the actual concentration of the solution

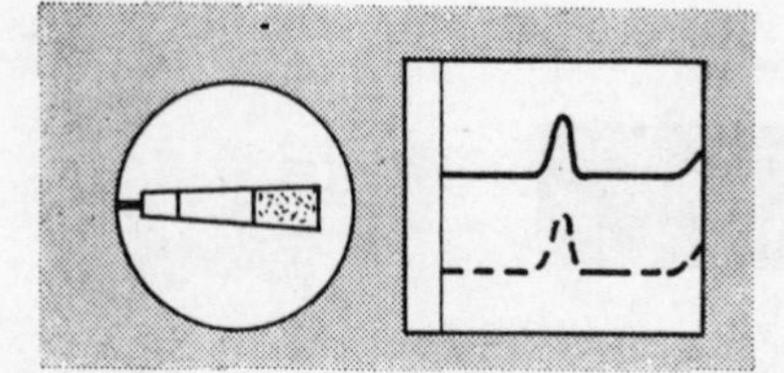

Wedge cells enable the picture to be displaced vertically from the normal position. A wedge cell can be run in the same rotor as a normal cell enabling two solutions to be compared under similar conditions

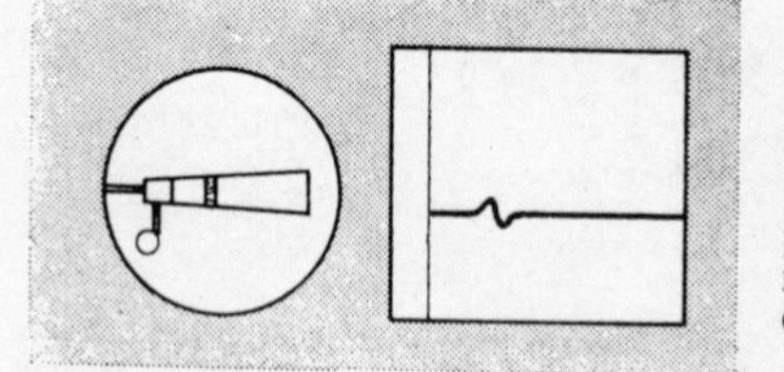

A band-forming centrepiece enables a solution to be layered over a denser buffer solution. Less material required, fast components are not running in slower materials because there is a physical separation. Schlieren shown here but may be used with ultraviolet absorption by choice

Figure 3.8. Some typical cells and the Schlieren pictures associated with them. Centrepieces are supplied in various materials depending on the chemical nature of the solution to be analysed

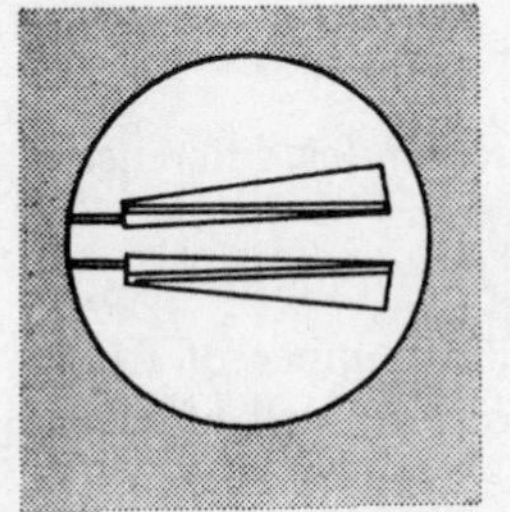

Interference cell has built in slits. Usually only shallow layer of liquid used to hasten attainment of equilibrium. Solvent column slightly overlaps solution column

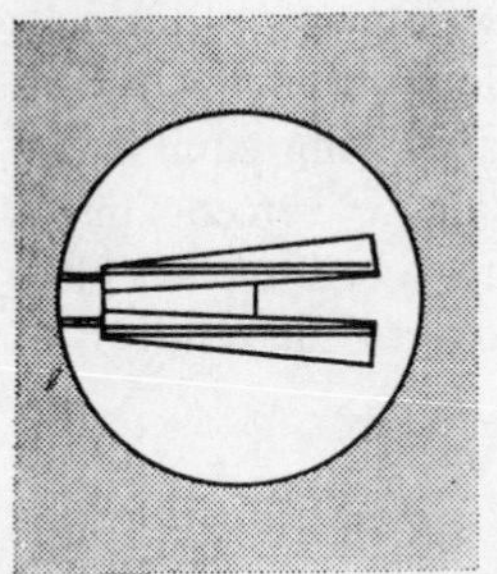

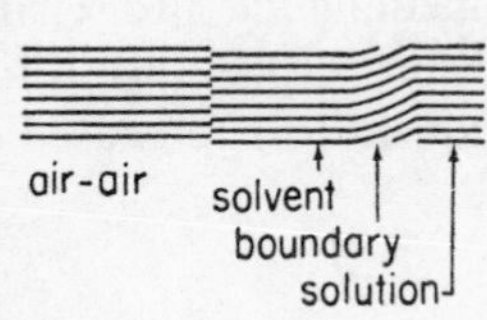

Synthetic boundary cell allows solvent layer at higher level to flow over through capillary on to solution under gravity. By counting fringes displaced across boundary formed δj obtained is direct measure of solute concentration

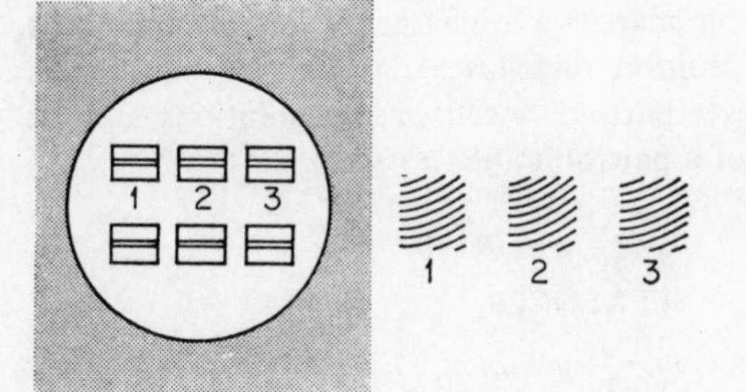

Multichannel centrepiece for simultaneous study of three solvent–solution pairs. Cavities have to be filled with cell partly dismantled before fixing upper window

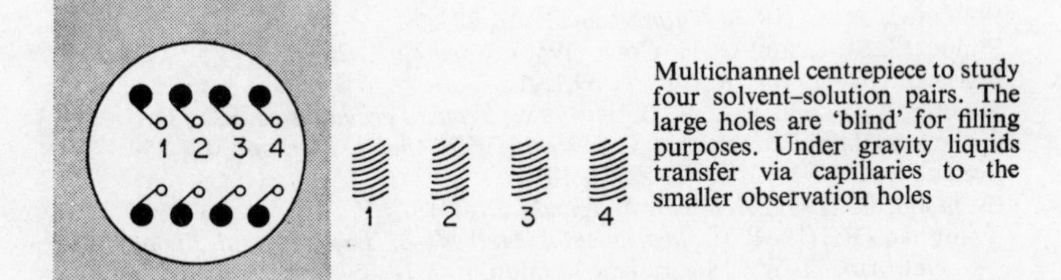

Multichannel centrepiece to study four solvent–solution pairs. The large holes are 'blind' for filling purposes. Under gravity liquids transfer via capillaries to the smaller observation holes

Figure 3.9. Some examples of interference cells. The fringes observed are all parallel and represent the change in refractive index (and therefore concentration) as a function of radial distance. The multichannel cells owe much to Yphantis (1960) for their development

slit. The MSE ultracentrifuge produces more fringes with some gain in convenience of measurement.

If a is the optical pathlength and there is a refractive index difference of $n_s - n_0$ between solution and solvent channels, then:

$$j \cdot \lambda = a(n_s - n_0) \tag{3.4}$$

where λ is the wavelength of the light and j is the number of fringes displaced. In practice, a 1 per cent protein solution in a cell of 12 mm pathlength gives a displacement of around 40 fringes.

In order to get more runs done in the same rotor it is possible to use a rotor with several cell holes. By using wedge windows in some of the cells it is possible to stack extra pictures above and below the one shown in Figure 3.7. When using the Rayleigh optics for fairly long equilibrium runs this method provides something like five separate experiments in the same rotor. However, when using wedge cells in this way the phaseplate has to be removed and the camera lens refocused due to the removal of the material of the phaseplate.

Special cells

Several cells are available for special purposes. Figure 3.8 summarizes some of the cell types and the kind of Schlieren patterns that would result from their use. Figure 3.9 summarizes the characteristics of some cells for interference work. In addition there are cells for the mechanical separation of upper and lower parts of a cell so that solutions may be withdrawn from both sides of a partition after a run.

REFERENCES

Aten, J. B. T. and J. A. Cohen (1965) *J. mol. Biol.*, **12,** 537.
Lamm, O. (1937) *Nova Acta R. Soc. Scient. upsal.*, **10,** No. 6, Ser IV.
Lee Gropper (1964) *Analyt. Biochem.*, **7,** 401.
Longsworth, L. G. (1943) *J. Am. chem. Soc.*, **65,** 1755.
Philpot, J. St. L. (1938) *Nature Lond.*, **141,** 283.
Philpot, J. St. L. and G. H. Cook (1948) *Research*, **1,** 234.
Rayleigh (1896) *Proc. R. Soc. A*, **59,** 201.
Shooter, K. V. and J. A. V. Butler (1956) *Trans. Faraday Soc.*, **52,** 734.
Spragg, S. P., S. Travers and T. Saxton (1965) *Analyt. Biochem.*, **12,** 259.
Svensson, H. (1939) *Kolloid.-Z*, **87,** 181.
Svensson, H. (1951) *Acta Chem. Scand.*, **5,** 1301.
Trautman, R. (1964) in *Instrumental Methods of Experimental Biology* (Ed. Newman, D. W.) Macmillan, London, p. 211.
Trautman, R. and V. W. Burns (1954) *Biochim. Biophys. Acta*, **14,** 26.
Wolter, H. (1950) *Ann. Physik*, **7,** 182.
Yphantis, D. A. (1960) *Ann. N.Y. Acad. Sci.*, **88,** 586.

CHAPTER 4

The importance of *s* and *D*

The sedimentation coefficient, s, and the diffusion coefficient, D, enter into the Svedberg equation for the determination of the molecular weight in Equation (1.11). The ratio of s to D can only be used for this purpose when they have both been determined in similar solvents and temperatures and corrected to infinite dilution. When molecular weights are determined by the Archibald method or by equilibrium methods this ratio does not appear as such in the equations used, because ways have been found to determine the ratio from a single experiment in the ultracentrifuge. However, when the density of the solute exceeds that of the solvent, sedimentation is a process always favouring the movement of solute to the bottom of the cell while diffusion is an opposing process that attempts to redistribute the solute evenly through the cell. Thus sedimentation and diffusion are working in opposition to each other, the magnitude of the centrifugal field deciding which shall be dominant.

The sedimentation coefficient

The sedimentation coefficient is the parameter of the solute that is determined most frequently. The experimental procedure for its measurement will be left until some of its general properties have been discussed. Schachman (1959) has discussed this at length. In the introductory account given here only the more salient points are dealt with.

The sedimentation coefficient, s, is the velocity of the solute molecules divided by the centrifugal field, hence:

$$s = \frac{\mathrm{d}x/\mathrm{d}t}{x\omega^2} = \frac{\mathrm{d}\ln x}{\omega^2\,\mathrm{d}t} = \frac{2{\cdot}303\log_{10} x}{\omega^2\,\mathrm{d}t} \tag{4.1}$$

provided that s is independent of x.
Thus, a plot of the logarithm of the distance of a 'peak' in the Schlieren pattern from the centre of rotation when plotted against time should give a straight line from the slope of which s (in seconds) is obtained. A Svedberg unit of $1S$ is equivalent to $s = 10^{-13}$ seconds.

Concentration dependence of s

The sedimentation coefficient varies with the concentration of solute and usually has to be corrected to infinite dilution to obtain a value of s^0. A plot of s against concentration, c, can be expressed over a *small* concentration range by:

$$s = \frac{s^0}{1+k_s c} \tag{4.2}$$

or by:

$$s = s^0(1-k_s c) \tag{4.3}$$

In these equations k_s is a constant. The linear form in Equation (4.3) has sometimes been used for theoretical work since it leads to simpler mathematics. An important consideration arising from these equations is that peaks are self-sharpening, since the tail of the peak is in a region of lower concentration than the leading edge. Some materials such as nucleic acids can be self-sharpening to an extreme degree. These materials need to be extrapolated with care to obtain s^0 values. The sensitivity of ultraviolet absorption optics has permitted very low concentrations to be measured, but at these very low concentrations convection is very likely to give trouble.

Sedimentation coefficients usually decrease as the total protein concentration increases. In the cases where s values increase with total protein concentration there is a dynamic equilibrium operating between the components of the mixture; this is discussed later (see Chapter 8). The reason for the decrease of s with increasing c has been discussed at length by Schachman (1959). Explanations have centred around detailed considerations of buoyant densities, viscosity effects and the backward flow of liquid as the solute is sedimenting. A porous plug analogy (Fessler and Ogston, 1951; Ogston, 1953) considers sedimentation to be analogous to the movement of a fluid through a stationary porous plug; however, the mathematical relationships deduced from this are not valid at high dilutions. Although these considerations are extremely fascinating they cannot be developed further in an introductory text book.

Charge effects

Charge effects can alter the measured values of s, depending on the composition of the salts present in the solvent. A protein, or any other biological colloid, cannot be rendered free of small ions by prolonged dialysis against distilled water. In such cases, small ions (H^+ or OH^-)

must always remain behind to balance the charge of the protein. An attempt to centrifuge this dialysed protein would cause a partial separation of the protein from its counter-ions, which in turn would lag behind it and exercise an electrostatic drag, known as the sedimentation potential. This sedimentation potential has an electrophoretic effect opposing the sedimentation. If no other ions are present it can be shown that the molecular weight determined will be too small by a factor $1/(1+n)$ where n is the number of counter-ions per macromolecule. Incidentally, in diffusion studies the counter-ions try to drag the protein along faster since the small counter-ions will diffuse faster than the protein. Therefore, in an ultracentrifuge the counter-ions work in opposite senses for sedimentation and diffusion, so that as a result the effects do not cancel out. The effect just described is known as a primary charge effect. In practice, the protein is analysed in the presence of a swamping excess of a neutral electrolyte such as potassium chloride; usually 0·1 M KCl is sufficient and, of course, all s values should be extrapolated to zero concentration. The effect of increasing the salt concentration is to shrink the ionic atmosphere more closely around the macromolecule where it can screen the charges more effectively. Thus at low salt concentrations the average position of the oppositely charged ionic atmosphere is further from the central macromolecule than it is at high salt concentrations.

A secondary charge effect is related to differences in the sedimentation behaviour of the different ions present in the salts used. In general, under a gravitational field, a salt such as KCl will redistribute itself as a whole with negligible separation of the positive and negative ions. Salts in which the anions and cations have different sedimentation behaviour are avoided since the separation of the positive and negative ions under centrifugation would produce a sedimentation potential affecting the movement of the macromolecule; this is a secondary charge effect. When a biocolloid must be handled in the presence of a buffer solution, for example phosphate, it is usually desirable to arrange that only about 0·05 M is contributed to the salt concentration by the buffer but that an additional 0·1 M KCl is present. The undesirable secondary charge effect of the buffer is thus kept to a minimum.

It is usually assumed that the solvent plus added salts can be considered as a single component. The validity of this assumption is of some interest. Creeth and Pain (1967) have discussed the question of charge effects recently in a review of methods available for the determination of molecular weights. Cassasa and Eisenberg (1960, 1961, 1964) showed that so-called anhydrous molecular weights can be obtained in three-component systems without explicit definition of the molecular interactions involved, providing that the solution is dialysed to equilibrium. It appears that all theoretical difficulties due to interactions in a three-component system can be completely avoided. The salts used for the third component should be of low molecular weight and present in sufficiently high concentrations to avoid very large Donnan effects. The Donnan effect is the unequal concentration of salts each side of a semipermeable membrane when a charged macromolecule, such as a protein, is on one side.

Corrections for s

It is customary to express the sedimentation coefficient in a corrected form, $s_{20,w}$, which is the value of the coefficient when pure water at 20°C is used as a solvent. The correction takes the form of density and viscosity corrections according to the following equation:

$$s_{20,w} = s\frac{\eta}{\eta_{20,w}} \cdot \frac{(1-\bar{v}\rho)_{20,w}}{(1-\bar{v}\rho)} \tag{4.4}$$

The viscosity terms have the larger effect and are easier to determine than the density terms. A simplified form of the equation is often used:

$$s_{20,w} = s\frac{\eta}{\eta_{20,w}} \cdot \frac{\rho_p-\rho_{20,w}}{\rho_p-\rho} \text{ (approximately)} \tag{4.4a}$$

where ρ_p is the buoyant density of the particle.
It is probably not safe to correct over a wide difference in temperature, but with modern apparatus it is not difficult to carry out sedimentation runs at 20°C. However, the biological material may change its conformation with temperature so that work at 20° may be undesirable. In Equation (4.4) the ratio of the viscosities is the most important term. The other part of Equation (4.4) involves the partial specific volumes, $\bar{v}$, and the density, ρ, of the solvent. The determination of $(1-\bar{v}\rho)$ is described in Chapter 5.

Although the sedimentation coefficient can be combined with the diffusion coefficient to determine molecular weights by the sedimentation–diffusion equation (1.11), the method is no longer very popular because of theoretical objections fully described by Creeth and Pain (1967). In general, a determination of the sedimentation coefficient is always made because it is easy to determine it when a velocity run is being undertaken to check the homogeneity of a preparation. The magnitude of *s* can suggest the size range in which the macromolecule lies and thus help in the planning of experiments for the determination of molecular weight. Most biochemists want to design preparative scale experiments, and a knowledge of the *s* values of the macromolecule and its associated components helps to put such plans on a scientific basis.

The experimental determination of s

Equation (4.1) requires that the value of x shall be determined at known times. Modern apparatus is provided with timers that automatically look after the regular recording of pictures. The value of ω may be checked by a directly-reading odometer on the Beckman instrument in which one unit corresponds to 6,400 revolutions of the rotor.* It is customary to include reference holes in the counterpoise (Figure 3.5) which have a known geometry relative to the centre of rotation. If another cell, such as a wedge cell, is being run in the rotor hole normally occupied by the counterpoise it is possible to unscrew plugs sealing reference holes drilled elsewhere in the rotor. In a typical run, using a standard cell and a counterpoise, a picture similar to that shown in Figure 4.1 will be obtained.

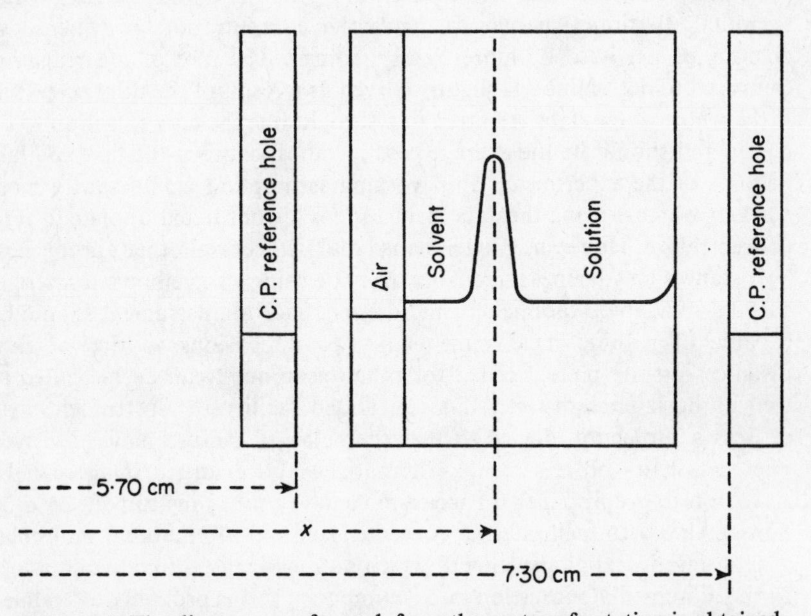

Figure 4.1. The distance, x, of a peak from the centre of rotation is obtained by interpolation between reference distances given by the apertures in the counterpoise. The actual dimensions refer to Beckman equipment

When the Schlieren peak is symmetrical it is customary to use its apex as the position of the boundary for the determination of x. Goldberg (1953) has proved that the square root of the second moment of the gradient

* The electronic speed control on newer instruments uses a different factor.

curve should be used. In practice the distinction is usually of little importance compared with the use of the centre of gravity.

Sedimentation occurs during the period of acceleration of the rotor, so it is a common practice to start the timing of the run when the rotor has reached $\frac{2}{3}$ of the maximum speed for the run in question. This rough correction assumes that the rotor has been accelerated uniformly. No difference is made to the value of s obtained by applying the correction, since a zero time correction does not alter the slope of log x against time. However, the straight lines plotted for a family of peaks in a mixture all converge to the meniscus position if a correct zero time correction is applied. Thus we may then add this point when determining s and it is particularly useful for the slower components of a mixture that are only resolved late in the run.

During sedimentation, the concentration of the solute decreases due to radial dilution (Chapter 5). Thus the concentration at which s is determined needs to be more closely defined. The plot of log x against time used to determine s is slightly curved as a result of the dilution effect. If the whole of the data are fitted to a straight line the concentration to be considered should be the average concentration between the first and last pictures of the experiment. This working assumption would satisfy most workers who are using the data from a series of runs to extrapolate to zero concentration. However, Schachman (1959) lists several other approaches.

Distances may be measured from the plate using a travelling microscope, such as an adapted toolmakers microscope. Such an instrument should be capable of making measurements in two dimensions so that vertical distances on the plate, needed for other techniques, can be measured as well as horizontal distances. As an added facility the instrument may possess a projection device so that the enlarged picture may be viewed over crosshairs on a screen.* Alternatively, an enlarged image can be traced on to graph paper for measurement. A refined instrument used by Spragg (1963) to facilitate the computer solution of elaborate molecular weight data is a digital scanner† which automatically prints the data in a form for immediate insertion into a computer; this is probably too refined for the simple calculation being described here. Obviously, interpolation between the reference points enables x to be determined for use in Equation (4.1). It is desirable to check the position of the air–solvent meniscus on the first and last photograph to make sure that no leak has taken place during the run.

* Such an instrument is available from Precision Grinding Ltd., Precision Works, Mill Green Road, Mitcham Junction, Surrey, England.

† 'Oscar', Benson and Lehner Ltd., West Quay Road, Southampton, England.

In order to ensure accuracy close to 1 per cent the temperature must be controlled to 0·25°C or less throughout the run, preferably at the temperature to which data are to be referred (usually 20°C) in order to avoid corrections using Equation (4.4) that may not be too reliable. The manufacturer's manual gives details of temperature measurement and control systems, usually determined by a thermistor within the rotor. Direct measurement of temperature is essential since a rotor cools due to adiabatic expansion when accelerated and warms up again when decelerated at the end of the run. For work of high precision the stretching of the rotor under gravity is allowed for, since it affects the measured dimensions. Kegeles and Gutter (1951) have described a correction obtained from photographs taken at high and low speeds. However, the problem is not simple since the axis of rotation may shift. Rotor stretching might cause a shift of 0·02–0·04 cm (Schachman, 1959).

When s values are very low their measurement is facilitated by the use of a synthetic boundary cell to form a peak part of the way down the cell, otherwise it might not be resolved from the meniscus.

(a) *Band centrifugation*

When centrifugation is carried out as described above the peaks observed using Schlieren optics correspond to the interfaces between solvent and solution. Leading peaks run in an environment made up of the slower components present (see Figure 3.1). In biochemical mixtures some of these slower components may be quite viscous as, for example, ribosomes are during analysis in a bacterial extract. Any uncorrected s values quoted may be considerably different from the $s^0_{20,w}$ values. Furthermore, when s values are obtained they may not bear a direct relationship to the values needed for preparative work carried out in density gradients. Such preparative runs are zonal in that the mixture to be analysed is layered on the top of the gradient (which prevents convectional disturbances) and the different components descend through the gradient physically separated from each other. A band-forming centrepiece may be employed which uses a layering technique similar to that encountered in zonal preparative work. Figure 3.8 shows one form of band-forming centrepiece (Beckman Instruments Inc.). Vinograd and Bruner (1966) have described the advantages of this kind of cell: less material is required, resolved components are physically separated, slow contaminants remain behind the moving bands and the solvent for sedimentation may be arbitrarily selected without recourse to dialysis. The sample solution has to be less dense than the sedimentation solvent and this is arranged by the choice of concentrations of the low molecular weight solutes. In the Beckman cell, 15 μl of sample is placed in the reservoir and layers on to the solvent when the rotor reaches a speed of 5,000–8,000 rpm. The liquid column is homogeneous at the start of the run. As the zones descend through the solvent they broaden due to the combined effect of diffusion and the increase of the gravitational field with radial distance.

In such a cell sedimentation coefficients can be determined by plotting $\log x$ against time as described above (x is the radial distance of the maximum in the concentration distribution). The s values obtained compare well with those obtained by boundary centrifugation (Vinograd, Bruner, Kent and Weigle, 1963).

Strohmaier (1966) has described a sector-shaped insert for use in a swing-out preparative rotor to facilitate the determination of s values for biological materials.

(b) *Determination of s from density gradients*

Methods for the determination of s during experiments using density gradients have been listed by Charlwood (1966). These can prove useful in connection with experiments using swinging bucket rotors. Let us assume that the particles are spheres and that Equations (1.2) and (1.3) are valid. By combining these equations we have the following:

$$\tfrac{4}{3}\pi r^3 \cdot x\omega^2(\rho_p - \rho) = 6\pi\eta r \frac{\mathrm{d}x}{\mathrm{d}t}$$

which simplifies to:

$$v = \frac{\mathrm{d}x}{\mathrm{d}t} = \tfrac{2}{9}r^2 \cdot \frac{x}{\eta}\omega^2 \cdot (\rho_p - \rho) \tag{4.5}$$

where v is the velocity of sedimentation, r the radius of the spherical particle, η the viscosity of the solvent, and ρ_p and ρ are the densities of particle and solvent respectively. If experimental conditions are arranged so that the buoyancy term $(\rho_p - \rho)$ is nearly independent of position (this only applies to proteins if the sucrose used to make the density gradient is not too concentrated) and the gradient is arranged so that η is linear with x, then the velocity, $\mathrm{d}x/\mathrm{d}t$ is going to be nearly constant (de Duve and coworkers 1959). In this case the velocity can be determined from the movement of the zone and a sedimentation coefficient determined in the usual way described above.

Relative methods can be used to determine s when the above conditions cannot be realized. Martin and Ames (1961) used markers of known s

values such as enzymes, Stanworth, James and Squire (1961) used dye-labelled proteins while Charlwood (1963) used radioactively labelled proteins. If the comparison substance interferes it can be loaded on a separate control tube.

In general, the determination of s from density gradients in preparative equipment cannot be compared in precision with data obtained from the analytical ultracentrifuge. However, there are occasions when the analytical machine may not be available, or when s values are needed under preparative conditions. For example, the material to be studied may only be available in radioactively labelled quantities (for example, animal viruses or RNA that has been obtained from them).

The diffusion coefficient

The diffusion coefficient, D, in which we are interested refers to translational motion; the coefficient of rotational diffusion will not be further discussed here. Gosting (1956) has reviewed the measurement and interpretation of diffusion coefficients in proteins and his article should be read for further detail. Early experiments on diffusion were hampered by a quantitative statement in terms of D until Fick (1855) defined it and expressed it in his first and second laws. Later, Stefan (1879) and Boltzmann (1894) provided integrated forms of Fick's laws that became the basis of practical methods for the determination of translational diffusion coefficients. With the development of the velocity method of determining molecular weights (Equation (1.11)) the determination of D for biological macromolecules became very important. Nowadays, with the increased employment of rapid techniques for using the equilibrium method for the determination of molecular weight the necessity to measure D is less pressing. However, the spread of a peak on sedimentation is closely concerned with the value of D for the relevant component, but we shall see that there are other factors to be considered as well.

Fick's law of diffusion

Fick's first law of diffusion may be stated in the form:

$$\mathrm{d}m = -DA\frac{\mathrm{d}c}{\mathrm{d}x}\,.\,\mathrm{d}t \tag{4.6}$$

where $\mathrm{d}m$ is the mass diffusing across an area, A, when there is a gradient of concentration, $\mathrm{d}c/\mathrm{d}x$, for a short time interval, $\mathrm{d}t$. This equation, originally empirical, can also be derived from first principles. Considered as an empirical statement it looks like a common-sense result that dm should be proportional to time, area and the concentration gradient, but one should beware of statements such as 'the concentration gradient is driving the solute across'; random movement of the molecules occurs all the time. In the micro sense, diffusion occurs among individual molecules even when the macro concentration has become evenly distributed through the cell. Volume changes on mixing are neglected, and indeed appear to be unimportant in studies on the diffusion of proteins.

If, in Equation (4.6), the units of mass used in dm and dc are the same, then D has the dimensions cm^2/sec. For most biological colloids D is of the order of 10^{-7} cm^2/sec. Equation 4.6 may be derived from first principles and at the same time can give a more precise meaning to D. The following argument is substantially that given by Alexander and Johnson (1949).

Einstein considered that diffusion pressure and osmotic pressure were identical in magnitude but opposite in direction. Let us consider the solute molecules inside a small element of volume of thickness dx, in a cylinder of cross-sectional area A. The number of molecules in this volume element $= NAc\,\mathrm{d}x$ where c is a molar concentration and N is Avogadro's number. From the van't Hoff equation for osmotic pressure:

$$\mathrm{d}\Pi = RT\,\mathrm{d}c \tag{4.7}$$

Hence total force acting $= -ART\,\mathrm{d}c$ (minus sign because force is opposite in direction for diffusion and osmosis).

$$\text{Force per molecule} = -\frac{ART}{NAc}\cdot\frac{\mathrm{d}c}{\mathrm{d}x} = -\frac{RT}{Nc}\cdot\frac{\mathrm{d}c}{\mathrm{d}x}$$

Now applying Equation (1.1) (Force = frictional coefficient × velocity):

$$-\frac{RT}{Nc}\cdot\frac{\mathrm{d}c}{\mathrm{d}x} = f\,.\,\text{velocity} \tag{4.8}$$

If we have dn moles of solute crossing the element of volume in time dt we have:

$$\mathrm{d}n = Ac\,.\,\text{velocity}\,.\,\mathrm{d}t \tag{4.9}$$

Hence, combining (4.8) and (4.9) to eliminate the velocity:

$$\mathrm{d}n = -\frac{RT}{Nf}\,.\,A\,.\,\frac{\mathrm{d}c}{\mathrm{d}x}\,.\,\mathrm{d}t \tag{4.10}$$

Equation (4.10) is a more precise statement of Fick's first law (Equation (4.6)) in which we have D defined in terms of the frictional coefficient f:

$$D = \frac{RT}{Nf} \tag{4.11}$$

Thus from (4.11) we can use D to find f, or, from Equation (1.6), one could arrive at f from sedimentation data.

Equation (4.11) is only valid for dilute solutions. It is also possible to arrive at D through non-equilibrium system thermodynamics. This derivation is more comprehensive than the one just given and leads to:

$$D = \frac{RT}{Nf_1}\left[1+c_1\frac{\mathrm{d}\ln\gamma_1}{\mathrm{d}c_1}\right] \tag{4.12}$$

Whereas Equation (4.11) applied to a dilute two-component system, Equation (4.12) is more general and involves γ_1, the solute activity coefficient (a function of the solute concentration c_1). Fick's equation (4.6) referred to the cell as the frame of reference, but the derivation of (4.12) involved the local centre of mass, and hence the f_1 from (4.12) is not strictly comparable with the f from (4.6). However, when results are extrapolated to infinite dilution Equations (4.11) and (4.12) become identical.

Fick's second law of diffusion can be derived simply from the first:

Consider an infinitesimal volume between planes at x and the volume element across the plane at x then the amount of solute leaving the volume element across the plane at $x+\mathrm{d}x$ is given by $\mathrm{d}m+\dfrac{\mathrm{d}(\mathrm{d}m)}{\mathrm{d}x}\,.\,\mathrm{d}x$

Hence the difference in amounts of solute entering and leaving the volume element is given by:

$$\mathrm{d}m-\left(\mathrm{d}m+\frac{\mathrm{d}(\mathrm{d}m)}{\mathrm{d}x}\mathrm{d}x\right) = -\frac{\mathrm{d}(\mathrm{d}m)}{\mathrm{d}x}\,.\,\mathrm{d}x$$

This change in amount occurred in a volume element with a volume of $A\,\mathrm{d}x$. Thus, we can write for the concentration change occurring in the volume element:

$$\frac{\mathrm{d}c}{\mathrm{d}t}\mathrm{d}t = -\frac{\mathrm{d}(\mathrm{d}m)/\mathrm{d}x}{A\,\mathrm{d}x}\mathrm{d}x = -\frac{1}{A}\,.\,\frac{\mathrm{d}(\mathrm{d}m)}{\mathrm{d}x} \tag{4.13}$$

If we differentiate (4.6) with respect to x and substitute in (4.13) we have Equation (4.14), which is a statement of Fick's second law of diffusion:

$$\frac{\mathrm{d}c}{\mathrm{d}t} = D\frac{\mathrm{d}^2c}{\mathrm{d}x^2} \tag{4.14}$$

It has been assumed that D does not vary with concentration. Two integrated forms of (4.14) are particularly valuable for the practical determination of D:

$$c = \frac{c_0}{2}\left(1-\frac{2}{\sqrt{\pi}}\int_0^y \exp(-y^2)\,\mathrm{d}y\right) \tag{4.15}$$

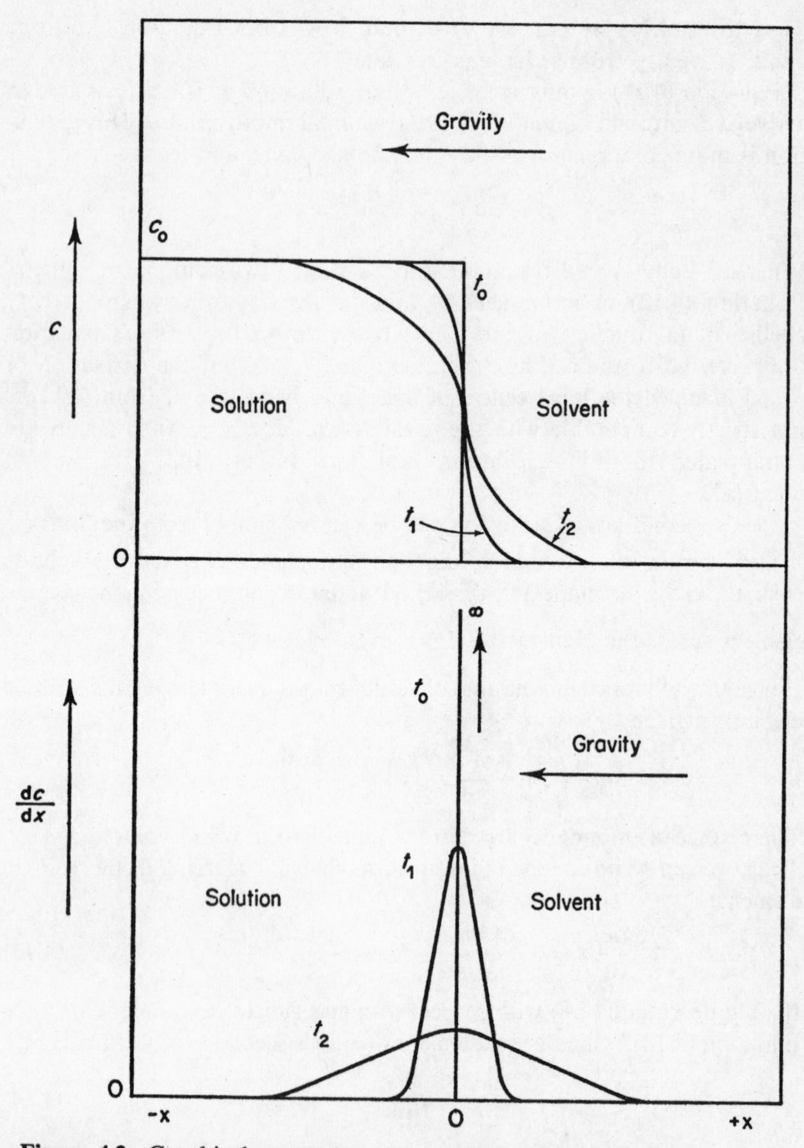

Figure 4.2. Graphical expression of Equations (4.15) above and (4.17) (below). In practice the solution is arranged to lie below the solvent to confer gravitational stability. An initially sharp interface broadens with time. In the lower graph the areas under the curves remain constant as the time (t_0, t_1, t_2) increases

where

$$y = \frac{x}{2\sqrt{(Dt)}} \tag{4.16}$$

The second term in the bracket of Equation (4.15) is the well-known probability integral found in statistical tables. The other integrated form of (4.14) is:

$$\frac{dc}{dx} = \frac{c_0}{2\sqrt{Dt}} \cdot \exp -\frac{x^2}{4Dt} \tag{4.17}$$

Figure 4.2 shows the graphical expression of Equations (4.15) and (4.17). The curves shown represent the kind of distribution to be expected from Fick's laws of diffusion. The experiment is usually arranged with the solution below the solvent so that gravity will stabilize the boundary but not be great enough to cause actual sedimentation. If sedimentation does occur, more complex relationships (to be described later) are required.

The experimentally determined value of D is usually converted to standard conditions such as zero concentration of protein in water at 20°C. This corrected and extrapolated value of D is then denoted $D^0_{20,w}$. However, the methods used for correction are not always beyond reproach. In particular, the temperature correction applied given by Equation (4.18):

$$D^0 = D^0_{T'} \frac{T\eta'}{T'\eta} \tag{4.18}$$

where T and T' are the two temperatures in °K and η and η' are the viscosities of the pure solvents at those temperatures. Experiments undertaken to test this equation have shown that it can only be used for comparing data at temperatures that are very close indeed. It is far better to make the measurements directly at 20°C.

It is customary to treat protein–buffer salt solutions as two-component systems, since the theory is neither sufficiently detailed nor convenient enough to allow for the flow interactions between all the components present. When extrapolating to infinite dilution this simplifying assumption appears to be a reasonable one. The correction of D^0 to its value D^0_w is made by using the relative viscosities:

$$D^0_w = D^0\left(\frac{\eta}{\eta_w}\right) \tag{4.19}$$

Equation (4.19) appears to convert the D value to pure water as solvent quite well provided that the change of environment does not produce aggregation or dissociation.

The experimental determination of D

The experimental determination of D is best carried out in a properly designed cell in static apparatus. However, a dynamic approach is possible using the ultracentrifuge providing that certain precautions and corrections are applied. In any case, an understanding of the factors involved in the spreading of peaks during an ultracentrifuge run is obligatory if sedimentation patterns are to be correctly interpreted.

(a) *Static methods*

Gosting (1956) describes various methods in great detail. As indicated above only one method of calculation of D has been considered here, the area–maximum height method. The use of the Schlieren optical system to follow the diffusion process does not give the highest precision but is the only method to be considered because it is part of the equipment of the analytical ultracentrifuge. Other optical systems that have been used for very precise work do not, in general, appear as standard equipment on an ultracentrifuge. A sharp interface is formed between solvent and solution so that the spread of the initially sharp boundary can be followed by a suitable optical system such as an interferometer, Lamm's scale method (Lamm, 1937), or a Schlieren optical system. Gosting (1956) should be consulted for detailed information, but here a fairly simple approach is intended so only the most popular method will be described. The area–maximum height method is very commonly used in conjunction with Schlieren optics. The lower curve in Figure 4.2 would be obtained using a Schlieren optical system. As the diffusion progresses the maximum height of the peak falls with time but the area should remain constant. After all, the area of the peak is a measure of the total concentration of solute across the boundary, which should remain constant throughout the run. If we make $x = 0$ at the centre of the boundary the exponential part of Equation (4.17) becomes unity and the equation simplifies to:

$$\frac{\mathrm{d}c}{\mathrm{d}x_{\max}} = \frac{c_0}{2\sqrt{(\pi D t)}}$$

Also, since concentration is directly proportional to refractive index (Equation (3.1)) we may write the following:

$$H_{\max}^2 = \frac{A^2}{4\pi D t} \qquad (4.20)$$

where A denotes area and $H_{\max}$ the height.

If there is an enlargement involved from cell to picture this has no effect on the ratio of H_{max} to A as long as allowance is made for the enlargement in the x ordinate. If D is not independent of concentration the peak will be skewed and a more involved calculation of the results will be needed. It has been assumed that a pure preparation is being measured. The shape of the resulting peak is then Gaussian, i.e. it can be defined precisely in terms of its mean and the standard deviation of its spread. If there is more than one component, the resultant curve from summing Gaussian peaks is no longer Gaussian. Theory is available for the solution of such resultant curves but is outside the scope of this book.

In practice it is not possible to obtain exactly sharp initial boundaries. If the blurring of the initial boundary is only slight one can correct the time scale by adding a zero time correction to allow for the slight bogus 'diffusion' that has occurred. According to Longsworth (1947), if the erroneous value of the diffusion coefficient is D' derived from Equation (4.20), if δt is the zero time correction and if the values of D' are obtained at experimental times t', then the true value of D can be found from the following equation:

$$D' = D\left(1+\frac{\delta t}{t'}\right) \qquad (4.21)$$

The value of $D\,\delta t$ should not exceed 10^{-4} cm^2. A graph of D' against $1/t'$ gives an intercept of D and has a slope of D.

The ideas so far expressed are reasonably valid for non-electrolytes, providing that extrapolation of the data to zero concentration is carried out. However, proteins are charged molecules and are handled in an environment of charged ions. If the diffusion coefficient of a protein that had been dialysed against distilled water were measured, it would be incorrect, since small ions would still be present. The protein would carry these small ions with it when diffusing in order to preserve electrical neutrality; hence the value of D obtained would not be merely a property of the protein but also of the small ions. The classical way round the problem is to measure D for the protein in the presence of a swamping excess of a neutral salt, such as 0·1–0·2 M KCl, against which the protein has been dialysed and which is used as the solvent side of the boundary. There are theoretical criticisms of the use of D so obtained in the determination of the molecular weight by the velocity method (described below); in particular, the effect of the salt environment on the partial specific volume, $\bar{v}$, is not fully clear. However, the use of a swamping excess of electrolyte is common practice with work on proteins.

A water bath is used to maintain constant temperature and should be arranged so that no vibration is transmitted to the cell by the stirrers. Temperature control to within $\pm 0{\cdot}001-0{\cdot}003$°C has been specified but Svensson (1951) has questioned the need for such high precision as a result of experiments in which temperature control was only to $\pm 0{\cdot}1$°C. The question depends on the viscosity and heat conductivity of the sample being studied. The protein should be dialysed to equilibrium against the buffer solution which will provide the solvent layer. Various cells have been devised to enable the solvent to be layered gently over the solution to form a sharp initial interface. The Lamm cell (Lamm, 1937) enables a slide to be withdrawn between solvent and solution compartments to form the boundary. The Neurath cell (Neurath, 1941) slides the solvent over on to the solution. Coulson, Cox, Ogston and Philpot (1948) devised a cell in which a sharp interface between solvent and solution was formed by a flowing junction technique. When the boundary was judged to be sharp enough the outflow from the cell was gently stopped and the timing of the diffusion run started. The firm of Zeiss have described their version of diffusion equipment (Reinert, 1967) using Philpot–Svensson Schlieren optics. The accuracy of the Schlieren optics was increased by the use of multipass cells which are traversed several times by the light through a non-mirror coated entrance and exit slit. The temperature control was to within 0·001°C per hour. Bugdahl (1967) has described another cell for use on the Zeiss Schlieren Equipment 80 which utilizes a teflon cone as a valve to allow solution to flow into the cell below the solvent so that the sharp initial boundary is formed by underlayering. The subsequent spreading of the boundary in this case was followed by Rayleigh–Svensson interferometry.

When Schlieren optics have been used, the employment of Equation (4.20) gives a plot of H_{max} against $1/\sqrt{t}$ and should produce a straight line of slope $A/2\sqrt{\pi D}$. The use of Equation (4.21) enables a zero time correction to be applied to the time values thus giving a straight line plot of Equation (4.20).

(b) *Determination of D in the ultracentrifuge*

We have already seen that sedimenting peaks are normally self-sharpening as a result of the concentration dependence of s (Equations (4.2) or (4.3)). This self-sharpening must be superimposed on the boundary spreading due to diffusion, and must be made negligible either by running the ultracentrifuge at low speed or by applying a correction for its effect at higher speeds. Furthermore, temperature control in an ultracentrifuge is not as good as the best standards available in static apparatus, but is claimed to be to within 0·1°C. Vibration is another problem and also rotor wobble at the lower speeds where self-sharpening becomes negligible. An extra-heavy rotor is marketed by Beckman Instruments Inc., for steady runs at low speeds, and MSE also claim very steady running at low speed in their instrument which has the motor in the rotor chamber. According to MSE, diffusion coefficients can be measured in their apparatus to an accuracy comparable with that obtained by static methods. For diffusion (and other) measurements MSE market a synthetic boundary cell with a light path of 20 mm. The synthetic boundary method for obtaining the initial sharp interface between solution and solvent may involve over-filling or under-filling, depending on whether the solvent is flowed over the solution or the solution is flowed under the solvent. A double-sector centrepiece is chosen when interference optics are to be used or to superimpose a true baseline under the diffusion peak when using Schlieren optics. Since the diffusion run is being carried out in a sector-shaped cell in the ultracentrifuge, the boundary will be slightly skew. However, it is still possible to use the area–maximum height approach since the height is almost unaffected. Skewness is also due to the variation of D with concentration but as a rule this effect is small. The commonest cause of skewness is due to a bad initial boundary and so a small zero time correction should always be aimed for.

At higher speeds the self-sharpening effect has to be corrected for. Schachman (1959) and Creeth and Pain (1967) give details of these corrections. An early correction due to Lamm (1929) was to use the following equation:

$$D = D_{\text{sed}}(1-\omega^2 st) \tag{4.22}$$

However, it was later shown by Fujita (1956) that a more involved treatment was required. Allowance was made for the variation of s with concentration according to a linear law (Equation (4.3)) and it was assumed that D was independent of concentration. This led to the equation:

$$\left(\frac{AF(\xi_m)}{H}\right)^2 = \frac{2D}{\omega^2 s^0}(\mathrm{e}^{\tau}-1)[1+(1-\lambda)^{\frac{1}{2}}]^2 \tag{4.23}$$

In this equation all the data on the right hand side are experimental except D. A is the area of the peak and H its maximum height. We assume $s = s^0(1-kc)$, $\tau = 2\omega^2 s^0 t$, and $\lambda = kc^0(1-\mathrm{e}^{-\tau})$,

$$\xi_m = \left[\frac{\omega^2 s^0 x_0^2 kc^0}{2D}\right]^{\frac{1}{2}}\left[\frac{\lambda^{\frac{1}{2}}}{1+(1-\lambda)^{\frac{1}{2}}}\right]$$

$$\Phi(\xi_m) = \frac{2}{\sqrt{\pi}}\int_0^{\xi_m} \mathrm{e}^{-y^2}\,\mathrm{d}y$$

$$F(\xi_m) = \frac{\Phi'(\xi_m)}{1+\Phi(\xi_m)}+2\xi_m \tag{4.24}$$

The data from these equations comes from experimental observations and the use of tables of probability functions. A plot of $AF(\xi_m)/H^2$ against $(e^{\tau}-1)[1+\sqrt{(1-\lambda)}]^2$ will be a straight line of slope $2D/\omega^2 s^0$. A value of D is assumed in the first place and then adjusted until the plot becomes linear, thus giving the correct D. The validity of this equation was tested by Baldwin (1957). Schachman (1959) has described the Fujita method very fully.

Van Holde (1960) used a simpler form of an expression given by Fujita (1959). This is simpler to use than the above treatment since it needs no ancillary data. However, it is more a test of polydispersity than a method of determining diffusion coefficients. Spreading due to diffusion varies as $t^{\frac{1}{2}}$ but spreading due to polydispersity varies directly with t so at infinite time the effect of diffusion would be negligible. The relevant equation is as follows:

$$\left(\frac{H}{A}\right)t^{\frac{1}{2}}(1-\tfrac{1}{2}\omega^2 st)^{-1} = (4\pi D)^{-\frac{1}{2}}+\left(1-\frac{2}{\pi}\right)\frac{x_m\omega^2 st^{\frac{1}{2}}}{4D}\left(\frac{k_s c^0}{1-k_s c^0}\right)(1-\omega^2 st)+\ldots \qquad (4.25)$$

Recently Kawahara (1969) has derived a still simpler form of Fujita's equation.

For simplicity, most measurements of D in the ultracentrifuge would be at the lower speeds to avoid the involved computation of the equations that must be applied at higher speeds. At higher speeds an analysis of peak spreading is necessary to detect difficult cases of heterogeneity rather than to determine D. Of course, applying a counsel of perfection, material should be purified as fully as possible before subjecting it to physical measurements.

In more concentrated solutions the effect of concentration dependence on s is more pronounced. It is possible to balance the self-sharpening effect against the spreading due to diffusion and attain a steady state (Creeth, 1964). It is possible to obtain an apparent diffusion coefficient for a single solute. If this apparent value is higher than that obtained by free diffusion at the same concentration, heterogeneity is indicated.

REFERENCES

Alexander, A. E. and P. Johnson (1949) *Colloid Science*, **1**. Oxford University Press, Oxford.

Baldwin, R. L. (1957) *Biochem. J.*, **65**, 503.

Boltzmann, L. (1894) *Annln. Phys.*, **53,** 959.
Bugdahl, V. (1967) *Jena Review*, **2,** 107.
Cassasa, E. F. and H. Eisenberg (1960) *J. phys. Chem.*, **64,** 753.
Cassasa, E. F. and H. Eisenberg (1961) *J. phys. Chem.*, **65,** 427.
Cassasa, E. F. and H. Eisenberg (1964) *Adv. Protein Chem.*, **19,** 287.
Charlwood, P. A. (1963) *Analyt. Biochem.*, **5,** 226.
Charlwood, P. A. (1966) *Br. med. Bull.*, **22,** 121.
Coulson, C. A., I. T. Cox, A. G. Ogston and J. St. L. Philpot (1948) *Proc. R. Soc. A*, **192,** 382.
Creeth, J. M. (1964) *Proc. R. Soc. A*, **282,** 403.
Creeth, J. M. (1964) *Proc. 6th Intern. Cong. Biochem., New York Abs.* (II), **39,** 146.
Creeth, J. M. and R. H. Pain (1967) *Prog. Biophys and mol. Biol.*, **17,** 217.
Duve, C. de, J. Berthet and H. Beaufay (1959) *Prog. Biophys. biophys. Chem.*, **9,** 325.
Fessler, J. H. and A. G. Ogston (1951) *Trans. Faraday Soc.*, **47,** 667.
Fick, A. (1855) *Annln. Phys.*, **94,** 59.
Fujita, H. (1956) *J. chem. Phys.*, **24,** 1084.
Fujita, H. (1959) *J. phys. Chem. Ithaca*, **63,** 1092.
Goldberg, R. J. (1953) *J. phys. Chem. Ithaca*, **57,** 194.
Gosting, L. J. (1956) *Adv. Protein Chem.*, **11,** 429.
Kawahara, K. (1969) *Biochemistry*, **8,** 2551.
Kegeles, G. and F. J. Gutter (1951) *J. Am. chem. Soc.*, **73,** 3770.
Lamm, O. (1929) *Z. Physik. Chem. A*, **143,** 177.
Lamm, O. (1937) *Nova Acta R. Soc. Scient. upsal. Ser. IV*, **10,** No. 6.
Longsworth, L. G. (1947) *J. Am. chem. Soc.*, **69,** 2510.
Martin, R. G. and B. N. Ames (1961) *J. biol. Chem.*, **236,** 1372.
Neurath, H. (1941) *Science*, **93,** 431.
Ogston, A. G. (1953) *Trans. Faraday Soc.*, **49,** 1481.
Reinert, K. E. (1967) *Jena Review*, **4,** 219.
Schachman, H. K. (1959) *Ultracentrifugation in Biochemistry*, Academic Press Inc., New York–London.
Spragg, S. P. (1963) *Nature, Lond.*, **200,** 1200.
Stanworth, D. R., K. James and J. R. Squire (1961) *Analyt. Biochem.*, **2,** 234.
Stefan, J. (1879) *Sber. Akad. Wiss. Wien* (2), **79,** 161.
Strohmaier, K. (1966) *Analyt. Biochem.*, **15,** 109.
Svensson, H. (1951) *Acta chem. scand.*, **5,** 72.
Van Holde, K. E. (1960) *J. phys. Chem. Ithaca*, **64,** 1582.
Vinograd, J. and R. Bruner (1966) *Fractions No. 1*, 2 Beckman Instruments Inc. Palo Alto, Calif., U.S.A.
Vinograd, J., R. Bruner, R. Kent and J. Weigle (1963) *Proc. natn. Acad. Sci. U.S.A.*, **49,** 902.

CHAPTER 5

A. The continuity equation and relationships derived from it

B. Partial specific volume

A. The Lamm equation

The continuity equation is also known as the Lamm equation (Lamm, 1929) or the general ultracentrifuge equation. It occupies a central position in the theory of sedimentation since it brings together the effects of s, D and ω into an equation relating concentration changes of solute with time. The sector angle of the cell is not ignored and the fact that sedimenting zones are curved. The centrifugal field is not homogeneous but increases linearly with x. The treatment leads to an understanding of the radial dilution law which helps correct Schlieren peak areas to actual concentrations. Expressions involving the molecular weight of the solute also follow from the Lamm equation.

Let us consider Figure 5.1, in which the various dimensions have been defined. The liquid in the cell is wedge-shaped with a curved meniscus and curved bottom and the sedimenting zones are also curved. The angle of the sector is θ radians so that the area of the centripetal face of the lamina at a radial distance x is given by θxh. It is possible to derive an equation for the change in concentration in the lamina of thickness dx if we consider the movement of solute through the centripetal and the centrifugal faces. This solute movement is composed of two components, namely sedimentation normally towards the bottom of the cell and diffusion in the opposite direction.

Thus, consider material carried across the plane at x by sedimentation invoking the definition of s (Equation (1.7)):

$$\frac{\mathrm{d}m_s}{\mathrm{d}t} = c\,.\,xh\theta\,.\,\frac{\mathrm{d}x}{\mathrm{d}t} = c\,.\,xh\theta\,.\,s\omega^2 x$$

In the opposite direction Fick's law (Equation (4.6)) applies to give:

$$\frac{dm_D}{dt} = -DA \cdot \frac{dc}{dx} = -D \cdot xh\theta \cdot \frac{dc}{dx}$$

Thus the net result of these two equations is to give the flow equation:

$$\frac{dm}{dt} = \theta xh\left(cs\omega^2 x - D\frac{dc}{dx}\right) \qquad (5.1)$$

The flow equation can lead to useful relationships for the determination of molecular weights and will be returned to later.

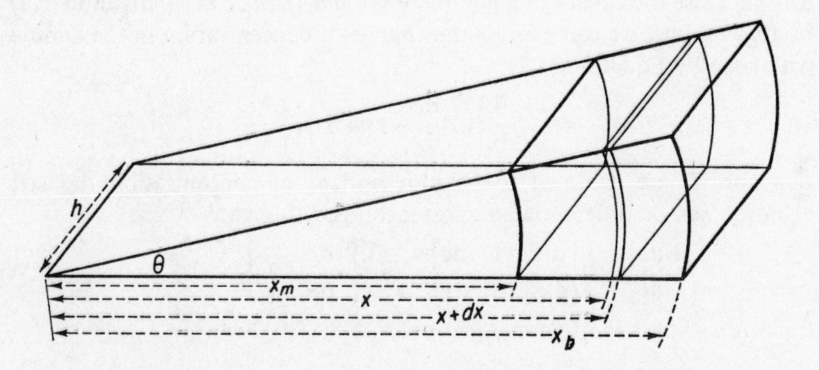

Figure 5.1. Illustration of a sector-shaped liquid in an analytical cell bounded by cell windows an optical distance h apart, by inclined cell walls at an angle θ to each other, and by the air–solution meniscus at radius x_m and cell bottom at radius x_b

An equation very similar to (5.1) can be established for the centrifugal plane of the lamina at $x+\mathrm{d}x$. This takes the form:

$$\frac{dm}{dt} = \theta h\left\{cs\omega^2(x+dx)^2 - D(x+dx)\left[\frac{dc}{dx}+\frac{d}{dx}\cdot\frac{dc}{dx}\cdot dx\right]\right\}$$

which, when expanded with the neglect of higher powers of small quantities, leads to:

$$\frac{dm}{dt} = \theta h\left\{cs\omega^2(x^2+2xdn) - D\left(x\frac{dc}{dx}+x\frac{d}{dx}\cdot\frac{dc}{dx}dx+\frac{dc}{dx}dx\right)\right\} \qquad (5.2)$$

When (5.2) is subtracted from (5.1) we have an expression for the net increase of solute in the volume element of thickness dx and volume θxh . dx as follows:

$$\left.\frac{\mathrm{d}m}{\mathrm{d}t}\right|_{\text{lamina}} = -\theta h\left\{cs\omega^2 2x - D\left(\underset{u}{x}\underset{\mathrm{d}v}{\frac{\mathrm{d}}{\mathrm{d}x}\cdot\frac{\mathrm{d}c}{\mathrm{d}x}} + \underset{v\,\mathrm{d}u}{\frac{\mathrm{d}c}{\mathrm{d}x}}\right)1\right\}\mathrm{d}x$$

This can be integrated, remembering that d(uv) = u dv + v du to yield Equation (5.3):

$$\left.\frac{\mathrm{d}m}{\mathrm{d}t}\right|_{\text{lamina}} = -\theta h\left\{\frac{\mathrm{d}}{\mathrm{d}x}\,.\,\underset{(uv)}{cs\omega^2x^2 - Dx\frac{\mathrm{d}c}{\mathrm{d}x}}\right\}\mathrm{d}x \tag{5.3}$$

This increase took place in a lamina of volume (θxh . dx) so, dividing (5.3) by this volume, we can express the change of concentration in the lamina with time by Equation (5.4):

$$\frac{\mathrm{d}c}{\mathrm{d}t} = \frac{1}{x}\cdot\frac{\mathrm{d}}{\mathrm{d}x}\left\{\left(D\frac{\mathrm{d}c}{\mathrm{d}x} - cs\omega^2x\right)x\right\} \tag{5.4}$$

If s and D are assumed to be independent of concentration the last equation can be differentiated and rearranged to give:

$$\frac{\mathrm{d}c}{\mathrm{d}t} = D\left\{\frac{\mathrm{d}^2c}{\mathrm{d}x^2} + \frac{1}{x}\cdot\frac{\mathrm{d}c}{\mathrm{d}x}\right\} - s\omega^2\left\{x\frac{\mathrm{d}c}{\mathrm{d}x} + 2c\right\} \tag{5.5}$$

Equation (5.5) is the Lamm equation.

The plateau region

In the plateau region dc/dx = 0 and d^2c/dx^2 = 0. Equation (5.5) then simplifies to:

$$\frac{\mathrm{d}c}{\mathrm{d}t} = -2\omega^2 sc$$

which, when integrated, leads to Equation (5.6):

$$c_t = c_0\,\mathrm{e}^{-2\omega^2 st} \tag{5.6}$$

Hence the concentration in the plateau region falls exponentially with time. The exponent in (5.6) is known as the equivalent time of centrifugation and is often given the symbol τ. Since it was assumed that s and D were constants in deriving the Lamm equation the use of Equation (5.6) is not strictly valid where s is a function of concentration. A more correct, but less convenient form, would then be:

$$c_t = c_0\,\mathrm{e}^{-2\omega^2\int_0^t s\,\mathrm{d}t} \tag{5.7}$$

Radial dilution

By combining (5.6) with the definition of s (Equation (1.7)) we have:

$$\frac{dc}{dt} = -\frac{2c}{x} \cdot \frac{dx}{dt}$$

which when integrated leads to the radial dilution equation below:

$$\frac{c_t}{c_0} = \frac{x_m^2}{x_t^2} \tag{5.8}$$

Thus, if the initial concentration of solute is c_0 before the Schlieren peak has left the meniscus at radius x_m the concentration falls to c_t when the peak has reached a radial position x_t. The particles increase their separation from each other since the gravitational field increases linearly along the radii. The combination of a linear law, due to the sector shape of the cell, and the linear law for the increase in gravitational field results in the square power relationship for the radial dilution correction.

Trautman and Schumaker (1954) have shown that it is possible to use Equation (5.8) even if s varies with c as a result of sectorial dilution.

Sedimentation equilibrium

Consider the flow Equation 5.1:

$$\frac{dm}{dt} = \theta xh\left(cs\omega^2 x - D\frac{dc}{dx}\right) \tag{5.9}$$

If centrifugation is carried on for a long time at a relatively low speed a steady state is finally reached so that sedimentation and diffusion have come to a state of equilibrium. The picture obtained is static and $dm/dt = 0$ at all points of the cell so that Equation (5.9) becomes:

$$\frac{s}{D} = \frac{dc/dx}{\omega^2 cx} \tag{5.10}$$

This applies at the equilibrium state no matter how s and D may depend on concentration (Archibald, 1963).

If Equation (5.10) is substituted in the Svedberg equation (1.11) we get one form of the equation for the determination of molecular weights by the method of sedimentation equilibrium:

$$M = \frac{RT}{(1-\bar{v}\rho)} \frac{dc/dx}{\omega^2 xc} \tag{5.11}$$

or

$$M = \frac{2RT}{(1-\bar{v}\rho)} \frac{\mathrm{d} \ln c}{\omega^2 \, \mathrm{d}x^2} \tag{5.12}$$

Thus a plot of the logarithm of concentration against the square of the radial distance should give a straight line, with the slope directly related to M.

At this point it is necessary to stress immediately that the above derivation of Equation (5.12) although simple, does not satisfy detailed analysis. The equation is a satisfactory basis for the determination of molecular weights but may be derived in a more rigorous fashion. More critical texts should be consulted by readers wishing to savour the subtleties of the subject. (Schachman, 1959: Fujita, 1962; Creeth and Pain, 1967). In the derivation of the Svedberg equation itself, the rigid definition of the nature of the density (relating to solvent or solution) is in some doubt, although at infinite dilution the question becomes of little practical importance since the data must always be extrapolated. The validity of Stokes' laws when applied to protein solutions is in question. The modern approach is to use non-equilibrium thermodynamics. In this connection the name of Hooyman (1953, 1956) is closely linked with more satisfactory, but difficult, derivations of equations that are identical with those obtained by simpler methods, providing an extrapolation of the data to low concentrations is made.

The general principles of irreversible thermodynamics allow us to deduce equations which are very similar in form to those derived above. These equations contain terms such as s, D, the activity coefficients, and so on. In order that these may be related to (for example) the molecular weight of a protein in a salt solution certain assumptions must be made. It is the particular virtue of the thermodynamic derivation that it clarifies for us exactly what these assumptions are. Thus, to give a specific example, it enables us to say that what was previously thought to be a necessary assumption in the derivation of the Svedberg equation, namely that the physical frictional coefficient involved in sedimentation and diffusion processes is equivalent, is in fact a superfluous assumption (see p. 152).

Approach-to-equilibrium

It is possible to derive molecular weights by the application of Equation (5.11) during the transient state of approach-to-equilibrium. Archibald (1947) showed that the equilibrium equation must also apply to the meniscus at the air–solution interface and at the cell bottom since there can be no transport of solute across these planes and dm/dt must equal zero. We can, therefore write Equation (5.11) in the more precisely defined form

$$M = \frac{R_T}{(1-\bar{v}\rho)} \cdot \frac{(dc/dx)_m}{\omega^2 x_m c_m} = \frac{RT}{(1-\bar{v}\rho)} \cdot \frac{(dc/dx)_b}{\omega^2 x_b c_b} \quad (5.13)$$

The way in which the above equations can be applied in practice will be dealt with in detail in the next chapter where the determination of molecular weights will be considered in some detail. The basic equations that have already been given for the three main methods of sedimentation–diffusion, equilibrium and transient state have several variants depending on the practical and theoretical problems that are likely to be encountered.

Solutions of the Lamm equation

Fujita and MacCosham (1959) have solved the Lamm equation in the following forms (Equations (5.14) and (5.15)):

$$\frac{dc}{dx} = \frac{c_0}{x_m} \frac{e^{-\tau}}{\varepsilon} \left\{ \left[1 - \Phi\left(\frac{z+\tau}{2(\varepsilon\tau)^{\frac{1}{2}}} \right) \right] \left(2 + \frac{z+\tau}{\varepsilon} \right) \exp \frac{z}{\varepsilon} - \frac{2}{\sqrt{\pi}} \cdot \frac{\varepsilon^{\frac{1}{2}}}{\tau} \cdot \exp - \left[\frac{(\tau - z)^2}{4\varepsilon\tau} \right] \right\} \quad (5.14)$$

where $\tau = 2\omega^2 st$ (ω = angular velocity of rotor and t = time) $\varepsilon = 2D/x_m^2\omega^2 s$, $z = 21 \log x/x_m$ where x is distance from the centre of the measured dc/dx. $\Phi(q)$ is the error function as described in Equations (4.24) and (4.15). Another integrated form is as follows:

$$\frac{c}{c_0} = \frac{e^{-\varepsilon T}}{2} \left\{ \left[1 - \Phi\left(\frac{T-z}{2T^{\frac{1}{2}}} \right) \right] - \frac{2}{\pi} \cdot T^{\frac{1}{2}} \cdot \exp - \frac{(T-z)^2}{4T} + [1+z+T] \left[1 - \Phi\left(\frac{T+z}{2T^{\frac{1}{2}}} \right) \exp z \right\} \quad (5.15)$$

where symbols are as for (5.14) except for $T = \frac{4Dt}{\varepsilon^2 x_m^2}$ and $z = \frac{2}{\varepsilon} \ln \frac{x}{x_m}$.

Such solutions are not suitable for hand calculation but are increasingly of interest for computer solution. For example, Spragg (1963) has measured values of dc/dx at two values of x within the boundary region at a given time of sedimentation and substituted them in Equation (5.14). The constants τ and ε were found by solving the two simultaneous equations. Then from τ and ε the coefficients s and D (and hence the molecular weight M) were calculated. The computer was programmed to solve Equation (5.14) by iteration. By using a digital plate reader (mentioned in Chapter 3) that records data in a form ready for the computer only 1 minute was required for measurement and 6 seconds for the digital computation. A new generation of analytical ultracentrifuges may be designed along these lines, but Creeth and Pain (1967) have drawn attention to a more pressing need than ever for the correct adjustment of machinery, particularly the optical systems. LaBar (1966) has used Equation (5.15) for defining sedimentation at the meniscus as long as a plateau region $dc/dx = 0$ exists in the cell. Such conditions occur in the Archibald approach-to-equilibrium technique. LaBar was able to lay down conditions when a linear extrapolation of the Schlieren plot of dc/dx to the meniscus could be justified. He found that the necessary condition for linearity was that Dt should be approximately 10^{-2} cm^2 and $\varepsilon > 0{\cdot}5$.

B. Density and partial specific volume*

The buoyancy term $(1-\bar{v}\rho)$ must now receive some consideration. In Chapter 1 it was shown to arise as a consequence of the principle of Archimedes. One modern method for the assessment of the partial specific volume depends on the application of the sedimentation equilibrium technique in solvents of different densities. Now that Equations (5.11) and (5.12) have been presented in this connection, it seems appropriate to group together all the important methods for the derivation of data concerned with density.

Hooyman (1953, 1956) has shown that $\bar{v}$ refers to the solute and ρ is the density of the solution in equilibrium experiments. The Svedberg equation is also embraced in these definitions. Admittedly, the density of solution approaches that of the solvent at infinite dilution.

The density of the solution can be determined in conventional density bottles and presents no problems. However, the partial specific volume is less straightforward. The partial specific volume can be defined as the volume increase when 1 gram of the solute is added to an infinite volume of the solution. Or,

$$\bar{v} = \left(\frac{dV}{dw}\right) \tag{5.16}$$

It is equivalent to the reciprocal of the effective density of the solute when actually dissolved in the solution, as a result of which its conformation will probably not be the same as in the dry state or even in another solution. The main source of error lies in the absolute determination of concentration of biological materials such as proteins. Even under the most favourable conditions it is hardly possible to obtain a mass concentra-

* See p. 152 for problems peculiar to solvents containing guaridine hydrochloride.

tion of a protein solution to better than 1 per cent and in most cases it can be of the order of 2–3 per cent. This arises simply because drying to constant weight is a most unsatisfactory procedure for solutes as labile as proteins.

There are three main methods of determining $\bar{v}$ in common use:

(1) by calculation from the amino acid analysis,

(2) by densitometry,

(3) by sedimentation–equilibrium studies in solvents differing in density.

The calculation method (1) is the most popular but rather suspect on theoretical grounds. Method (2) is so tedious in practice and so demanding of raw material that it is frequently avoided. Method (3) is a promising micromethod that produces good results but not without making some assumptions that may not always be justified. It does, however, have particular merit in not requiring knowledge of the protein concentration. These methods will now be dealt with in more detail.

Method (1)

An amino acid analysis is usually available from an automatic amino acid analyser. The individual partial specific volumes of each amino acid are also available in the literature (Cohn and Edsall, 1941). An assumption that the partial specific volume of the protein is an additive property of the

Table 5.1

Amino acid	$\bar{v}$	*Per cent by weight*	*Product of $\bar{v}$ and wt per cent*
Glycine	0·64	5	3·2
Valine	0·86	20	17·2
Tyrosine	0·71	40	28·4
Lysine	0·82	5	4·1
Aspartic acid	0·60	30	18·0
		100	70·9

Then $\bar{v}$ for the protein = 70·9/100 = 0·709

partial specific volumes of the constitutive amino acids is then made to calculate an answer. Obviously, there must be no non-protein component present. The assumption of additivity of the partial specific volumes is doubtful. Since proteins are made up of similar amino acids it is not surprising that $\bar{v}$ is generally somewhere in the region of 0·725, a value which is sometimes used when no other data are available. An error of 1 per cent in $\bar{v}$ leads to an error of approximately 3 per cent in a molecular weight because it enters as $(1-\bar{v}\rho)$ in calculations.

We can make up a fictitious example of the method as shown in Table 5.1, making the improbable assumption that only five amino acids are present.

Method (2)

The factor $1-\bar{v}\rho$ is obtained by the application of the equation:

$$1-\bar{v}\rho = \frac{1-w}{m} \cdot \frac{dm}{dw} \tag{5.17}$$

where w = weight fraction of solute (equals weight of solute divided by sum of weights of solute and solvent), m = the mass of liquid in the pyknometer and ρ = the density of the solution. The derivation of (5.17) will not be given here but is usually given in textbooks of physical chemistry or thermodynamics.

This method simply calls for pyknometers, a thermostat bath, a good balance, lots of protein (comparatively speaking) and patience. The Cahn Instrument Co. Ltd.,* have developed an instrument that makes use of the principle of weighing a plummet immersed in the solution. It is claimed to possess an accuracy of 0·000025 g/ml and requires only 1 ml of liquid. The cell has three concentric chambers; the outer chamber has water circulated through it at constant temperature, the middle chamber contains solvent to saturate the air above the solution, which in turn is contained in the centre chamber and in which the plummet is weighed. The apparatus is used with a Cahn electrobalance.

Method (3)

Edelstein and Schachman (1967) have described a micromethod for the determination of M and $\bar{v}$ from two sedimentation equilibrium runs in solvents of different density. If we consider the equilibrium Equation 5.12:

$$M = \frac{2RT}{(1-\bar{v}\rho)} \frac{d \ln c}{\omega^2 \, dx^2} \tag{5.18}$$

it is possible to obtain M, if we know $\bar{v}$, from the slope of a plot of ln c against x^2. Alternatively, knowing M we could find $\bar{v}$. Since two equations

* Cahn Instrument Co. Ltd., 27 Essex Road, Dartford, Kent, England. 15505 Minnesota Avenue, Paramount, California, U.S.A.

are required if two unknowns have to be found, a second run could be carried out in D_2O. However, the D_2O brings in another quantity, k, which is a factor by which M is increased due to binding of heavy water and by which $\bar{v}$ is decreased for the same reason. Therefore, for the run in D_2O, Equation (5.12) becomes:

$$kM = \frac{2RT}{\left(1-\frac{\bar{v}}{k}\rho_{D_2O}\right)} \frac{d \ln c}{\omega^2 \, dx^2} \tag{5.19}$$

The density of heavy water is known, or can be determined for diluted solutions. The value of k can be estimated from a knowledge of the number of exchangeable hydrogen atoms in the solute molecules (Hvidt and Nielsen, 1966). For bovine serum albumin, $k = 1{\cdot}0155$ but it is considered to be relatively constant for all proteins. Obviously, this assumption raises a slight doubt in the mind but the method is, nevertheless, very valuable because it is simple, economical and quick (since the two runs can be carried out in the same rotor) and does not require protein concentration determinations. The solution of Equations (5.18) and (5.19) simultaneously to find both $\bar{v}$ and M presents little difficulty. Greater accuracy can be obtained by using the denser $^{18}D_2O$ for the second run. Values of $\bar{v}$ refer to unhydrated solute.

REFERENCES

Archibald, W. J. (1947) *J. phys. Colloid Chem.*, **51,** 1204.

Archibald, W. J. (1963) in *Ultracentrifugal Analysis in Theory and Experiment* (Ed. Williams, J. W.), Academic Press, New York–London, p. 39.

Cohn, E. J. and J. T. Edsall (1941) *Proteins, Amino Acids and Peptides*, Reinhold, New York.

Creeth, J. M. and R. H. Pain (1967) *Prog. Biophys. mol. Biol.*, **17,** Pergamon Press, Oxford.

Edelstein, S. J. and H. K. Schachman (1967) *J. biol. Chem.*, **242,** 306.

Fujita, H. (1962) *Mathematical Theory of Sedimentation Analysis*, Academic Press, New York–London.

Fujita, H. and F. J. MacCosham (1959) *J. chem. Phys.*, **30,** 291.

Hooyman, G. J. (1956) *Physica*, **22,** 751.

Hooyman, G. J., H. Holton, Jr., P. Mazur and S. R. de Groot (1953) *Physica*, **19,** 1095.

Hvidt, A. and S. O. Neilsen (1966) *Adv. Protein Chem.*, **21,** 287.

LaBar, F. E. (1966) *Biochemistry*, **5,** 2362.

Lamm, O. (1929) *Arkiv. Mat. Astron. Fys.*, 21B, No. 2.

Schachman, H. K. (1959) *Ultracentrifugation in Biochemistry*, Academic Press, New York–London.

Spragg, S. P. (1963) *Nature Lond.*, **200,** 1200.

Trautman, R. and V. N. Schumaker (1954) *J. chem. Phys.*, **22,** 551.

CHAPTER 6

The determination of molecular weights*

We are now in a position to measure the molecular weights over a very wide range, from a few hundred to several million daltons. The material given here should be supplemented by reading the review by Creeth and Pain (1967) while the article by van Holde (1967) is most helpful as a practical guide to sedimentation equilibrium. Chervenka (1969) has published a useful manual based on Beckman instrumentation.

First let us recapitulate. There are three main methods for the determination of molecular weight; Equations (1.11), (5.11), (5.12), and (5.13) may be restated here for convenience:

$$\text{Sedimentation–diffusion:}\quad M = \frac{RTs}{D(1-\bar{v}\rho)} \tag{6.1}$$

$$\text{Equilibrium:}\quad M = \frac{2RT\,\mathrm{d}\ln c}{(1-\bar{v}\rho)\omega^2\,\mathrm{d}x^2} = \frac{RT\,\mathrm{d}c/\mathrm{d}x}{(1-\bar{v}\rho)\omega^2 cx} \tag{6.2}$$

$$\text{Approach-to-equilibrium:}\quad M = \frac{RT}{(1-\bar{v}\rho)}\cdot\frac{(\mathrm{d}c/\mathrm{d}x)_m}{\omega^2 c_m x_m} \tag{6.3}$$

These equations may need to be transformed slightly depending on the technique being used. From the theory of non-equilibrium thermodynamics it appears that these equations require the right hand side to be multiplied by the following factor:

$$\text{Thermodynamic factor} = 1 + c\frac{\mathrm{d}\ln\gamma}{\mathrm{d}c} \tag{6.4}$$

This factor presents a difficulty in that the data for the activity coefficient, γ, are not easily accessible. However, at infinite dilution the factor becomes unity. The important point to remember is that only data at infinite dilution can be used in Equations (6.1), (6.2) and (6.3), or, at least, one

* See p. 151 for information on chain molecular weights.

should be satisfied that measurements have been made into regions where the factor becomes negligible. Compact molecules of low charge are likely to give the best results, but some molecules can pose special problems when they are highly asymmetric and highly charged, nucleic acids for example. Charge effects have to be suppressed by working in the presence of suitable salt such as 0·1 M KCl and it may be possible to reduce the charge of a protein by adjusting the pH nearer the isoelectric point. High speeds of rotation do not normally affect the frictional properties of macromolecules due to the assumption of preferred orientations but varying angular velocities should be tried since fresh information may be revealed. For example, if the data near the air–solution meniscus are being used, as in the Archibald method, varying speeds and times may show the presence of a lighter component in the sample. DNA poses particularly difficult problems and one approach to its molecular weight is via intrinsic viscosity data, a subject reserved for treatment in Chapter 7. Aten and Cohen (1965) should be consulted for more information on the problem of DNA.

Molecular weight averages

The molecular weights calculated for mixtures may be of various types, defined as follows:

$$\text{Number average } M_n = \frac{\Sigma n_i M_i}{\Sigma n_i} \tag{6.5}$$

$$\text{Weight average } M_w = \frac{\Sigma n_i M_i^2}{\Sigma n_i M_i} \tag{6.6}$$

$$z\text{-Average } M_z = \frac{\Sigma n_i M_i^3}{\Sigma n_i M_i^2} \tag{6.7}$$

In these equations n_i is the number of molecules with molecular weight M_i. The number average is the type of average commonly sought in everyday experience as, for example, when determining the average height of a group of people. In the molecular weight field a number average is given by colligative methods such as osmotic pressure measurements or by the determination of chain length by end-group analysis. The weight average is calculated using weight in the denominator and emphasizes the contribution made by the heavier components of a mixture. Weight averages are generally given by sedimentation equilibrium and also by light scattering estimates, in which one particle scatters as much light as 10,000 of 1/100th the size. As will be shown below a z-average result can arise when the results of sedimentation equilibrium are treated in a special way. Sedimentation–diffusion gives an average intermediate between a number and a weight average result. It is apparent that an understanding of the type of average given is of great importance when comparing results from different techniques used on the same material. Fortunately, all averages give the same answer if the material is pure. Conversely, different answers for M_n, M_w or M_z suggest heterogeneity. The biologist must therefore attempt to purify his material as highly as possible before determining its molecular weight. In Figure 6.1 a distribution of molecular weights is shown and the corresponding M_n, M_w and M_z results are indicated; the number average falls nearest the value for the major component present but only coincides with it when the distribution is symmetrical. Equation (8.12) relates M_w and M_z in two component systems.

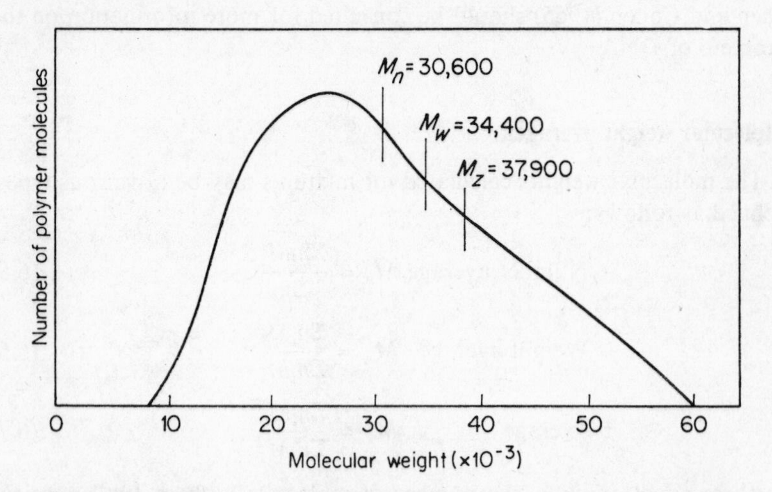

Figure 6.1. Frequency distribution of a mixture of polymers showing the calculated values of M_n, M_w and M_z

Machine representation of data

At the time of writing it is not possible to buy an ultracentrifuge that will present the molecular weight fully calculated from data processed in an internal computer. The temperature will have been measured using a thermistor device that has been precalibrated with the rotor outside the machine and with a thermometer in the cell hole of the rotor. The angular velocity will have been calculated from revolutions read off from a directly-driven odometer. The information on concentration (or refractive index) or its derivative will have been recorded on a photograph as a Schlieren or Rayleigh picture. Since information has been given earlier (Chapter 3) on methods of interpolating distances from the centre of rotation using plate measurements, the main point that must be made now relates to the measurement of Rayleigh fringes. If a double-beam scanner was used, the data will be presented as a plot of optical density against radial distance and will include reference data relating to the geometry of the counterbalance reference holes.

A Rayleigh interferogram appears as a family of parallel curves. The vertical displacement of these curves is proportional to the refractive index difference between the two channels (solute and solvent) of the interference cell. The number of fringes displaced is given by Equation (3.4). The corresponding concentration change would be given by Equation (3.1). Fortunately it is not usually necessary to apply these equations, since one can work directly in terms of the number of fringes displaced or even by using a direct measurement in terms of distance actually measured on the plate. This simplification can be made in equilibrium work, since in Equation (6.2) we are simply interested in differences of the logarithm of concentration; a little thought will show that all the proportionality constants involved in relating concentrations in the cell to fringe displacements will have no effect on a $d \ln c$ plot except to displace it without altering its slope. Figure 6.2 shows the types of Rayleigh plot that are usually encountered. In Figure 6.2(a) a synthetic boundary double sector cell has been used to obtain a calibration of the original concentration, c^0, in terms of a count of fringes displaced. Such a cell has fine interchannel grooves pressed into the centrepiece of the cell to allow solvent to flow across from one channel to form a layer over the solution in the other channel as the rotor is accelerated. This kind of run is often carried out in connection with the low speed equilibrium method when there is no meniscus depletion of solute. In Figure 6.2(b) an equilibrium run has been carried out at high speed so that no solute remains at the meniscus, while Figure 6.2(c) refers to a low speed equilibrium run that has not resulted in depletion of the solute at the meniscus. In equilibrium runs it is customary to allow the solvent channel to overlap the solution at either end very slightly. The simple salts in solvent and solution redistribute under gravity to a similar extent and are ignored by the Rayleigh optics which merely record the solute redistribution. In Plate VII the actual fringes photographed are shown corresponding to the drawings in Figure 6.2. The photograph shown in Plate VII(a) for the synthetic boundary run was taken late in order to clarify the appearance

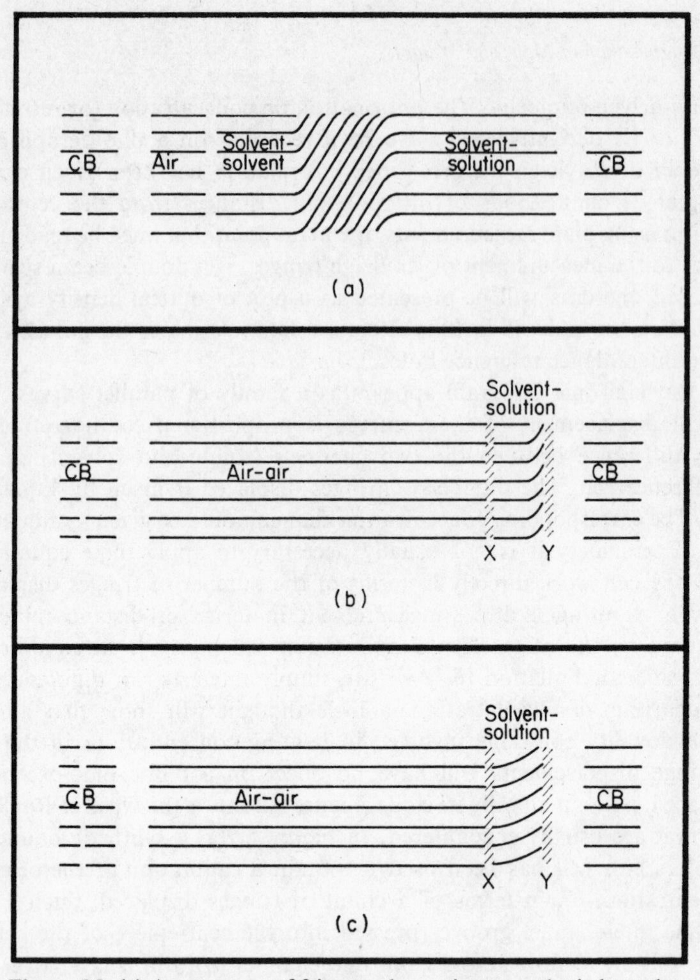

Figure 6.2. (a) Appearance of fringes when using a synthetic boundary cell. The number of fringes crossed from solvent to solution is proportional to the concentration of the solution.

(b) Appearance of fringes in a short path meniscus depletion equilibrium run. Reading along the fringe from X–Y we effectively have a set of parallel graphs of fringe displacement (proportional to concentration) against radial distance. The level portion of the fringe near X corresponds to zero solute concentration. The high speed Yphantis method produces such a picture.

(c) Appearance of fringes for a low speed equilibrium run. As in (b) the region X–Y effectively plots fringe shift (proportional to concentration) against radial distance. However, unlike (b) there is no meniscus depletion so the zero on the vertical concentration scale is not immediately apparent. In this case further calculation or experiment, is necessary to get the vertical concentration scale correct. CB denotes counterbalance references of known geometry (Chapter 3). The solvent path slightly overlaps the solution path at each end and a layer of fluorocarbon oil FC-43 at the bottom of the cell gives a correctly shaped meniscus in (b) and (c) of the fringes which in the earlier photographs would have been tightly spaced and changing position rapidly due to diffusion.

When measuring plates it is important to line up the air–air and counterbalance fringes parallel with the x-axis of the comparator. A run should be carried out using solvent against solvent to check that the fringes are still horizontal under the conditions of the experiment. However, cell windows can distort (although less so when sapphire windows are used), rotors can distort and cells can tilt slightly in the rotor at speed. Any slight distortion of the solvent–solvent fringes under these conditions should be allowed for when measuring the plates from the solvent–solution run.

Sedimentation–diffusion

The use of Equation (6.1) in conjunction with the values of s and D, measured and corrected as described in Chapter 4, leads to a value of the molecular weight. For example, if we consider bovine serum albumin and use the following data:

$D^0_{20,w} = 6{\cdot}45 \times 10^{-7}\ \mathrm{cm^2/sec}$

$\bar{v} = 0{\cdot}733$ calculated from the amino acid composition,

$\rho = 0{\cdot}9982$ for water at 20°C,

$s^0_{20,w} = 4{\cdot}73 \times 10^{-13}\ \mathrm{sec}$

then,

$$M = \frac{8{\cdot}314 \times 10^7 \times 298 \times 4{\cdot}73 \times 10^{-13}}{6{\cdot}45 \times 10^{-7}(1 - 0{\cdot}733 \times 0{\cdot}9982)} = 67{,}800.$$

This method is no longer very popular because it requires a separate experiment to determine D. At one time it was felt that this method was more useful with material that could not be produced in the degree of purity needed for equilibrium work. However, where the material is not pure there is uncertainty about the kind of average that will be obtained. Jullander (1946) showed that the result was somewhere between M_n and M_w; the exact position depends on the distribution of weights involved but it is probably nearer M_w. The method is not so economical in material as are the equilibrium methods. However, sedimentation–diffusion is often used as a cross-check on the molecular weight obtained otherwise.

Equilibrium methods

During a long run the concentration in the cell changes as shown in Figure 6.3. Initially all the solution was at the original concentration, c^0.

After a sufficiently long time a steady state is set up in which the concentration no longer changes at any point in the cell. At one point, the hinge point, there will always be a concentration equal to c^0. Under these conditions molecular weights may be obtained using Equation (6.2) or a variant of it obtained by some mathematical operations.

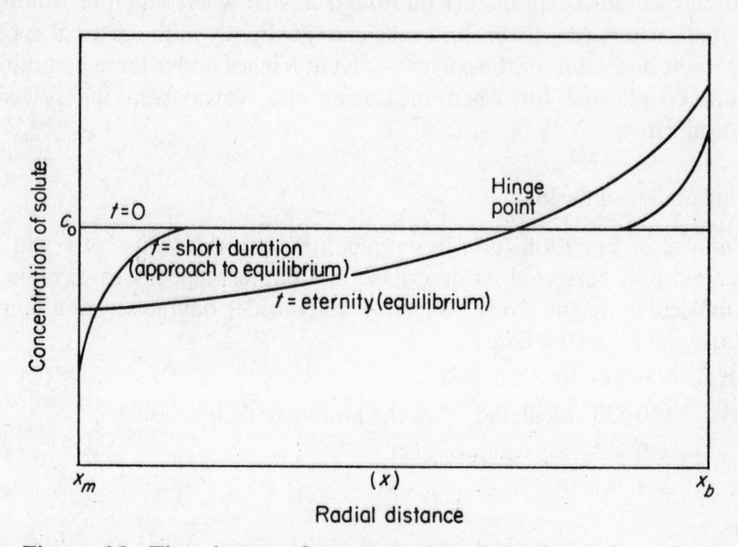

Figure 6.3. The change of concentration throughout the cell as equilibrium is approached

Equilibrium methods have a sound theoretical basis and can handle a wide range of molecular weights from a few hundred up to several millions. A major disadvantage used to be the long time required to reach equilibrium. However, Van Holde and Baldwin (1958) proposed the use of short columns of liquid from 1–3 mm high compared with the 14 mm or so for a completely filled cell. The time taken to reach equilibrium varies directly as the square of the height of solution so a great saving in time became possible. They found that ribonuclease reached equilibrium in 14 hours in a 3 mm column. Yphantis (1960) made a further study of the feasibility of using short columns. If dc/dx is measured at the hinge point corresponding to c^0 the molecular weight can be found directly. The hinge point is easily found with an ultraviolet double beam scanner, but in Rayleigh work there is a series of parallel fringes to choose from. The correct choice of fringe may be helped if the white light or zeroth fringe can be picked out. Otherwise, a swinging slit may be positioned over the plate, so that a great number of pictures may be taken on a single plate; this enables one to follow the movement of individual fringes as equilibrium is approached.

Van Holde and Baldwin (1958) showed that at a point x' given by the equation:

$$x' = \sqrt{\frac{(x_m^2 + x_b^2)}{2}} \tag{6.8}$$

which is within 0·002 mm of the midpoint of a 1 mm column, the concentration is given by:

$$c_{x'} = \frac{c^0 H}{\sinh H} = \frac{c^0}{1 + H^2/6 + \ldots} \tag{6.9}$$

where

$$H = \frac{\omega^2 M(1 - \bar{v}\rho)(x_b^2 - x_m^2)}{4RT} \tag{6.10}$$

For ribonuclease in a 0·7 mm column the concentration is within 1 per cent of c^0 for speeds under 24,000 rev/min or for a 3 mm column under 11,600 rev/min. Under these conditions it is only necessary to measure the dc/dx gradient at the midpoint of the cell and equilibrium is reached in 40 minutes. The advantages of using short paths are that it enables one to decrease the running times, to use higher speeds (Equation (6.10)) and hence more dilute solutions (Equation (6.2)), and also to use multichannel centrepieces (Figure 3.9). For equilibrium work the channels need not be sector-shaped, in fact, equilibrium can be reached faster when they are not since less solute may have to be transported to reach the equilibrium distribution. However, when using multichannel cells care has to be taken to avoid interchannel leaks due to scratches on the centrepiece and it is as well to try a dummy run with solvent pairs to check on distortion.

The high speed equilibrium method

Meniscus depletion or Yphantis method

In this method the ultracentrifuge is run fast enough for the fringes to become horizontal at the meniscus at equilibrium due to the depletion of the meniscus region of all solute (Figure 6.2(b)). Each fringe is a graph relating concentration (y-axis) to distance from centre of rotation (x-axis). As explained earlier, the y-axis is only proportional to concentration, but it can also be taken as a measure of solute concentration since factors disappear in the equations used. The fringes are level at the meniscus and thus the vertical concentration axis is characterized, since solute concentration is zero at the meniscus. Without a doubt, where it can be applied this approach is now one of the most popular for the determination of molecular weights because there is no need either for a second run in a synthetic boundary cell or for the further calculations needed in the other equilibrium methods. When the material is not pure the data nearest the meniscus refer to the lightest component, so that both this molecular weight and the weight average molecular weight for the mixture as a whole can be evaluated when all the data along the full liquid column is used.* However, the concentration range being studied can vary widely from zero concentration upwards, so this method might not be so desirable for strongly concentration-dependent systems.

The time taken to reach equilibrium is an exponential function and so, strictly speaking, an eternity is required to reach it. Although it was stated above that time taken to reach equilibrium varies as the square of the height of the solution column, this is not so for the meniscus depletion method, where the time taken varies linearly with the height of the solution column. Otherwise the time taken to reach equilibrium within 0·1 per cent is given by:

$$t_{0,1} = 0{\cdot}7\frac{(x_b - x_m)^2}{D} \tag{6.11}$$

where D is the diffusion coefficient; this equation is approximate. As a check on the proximity of equilibrium, one can take a series of photographs until the fringes show no further change in position.

An example of this method follows. An Yphantis (1964) six-place centrepiece was used (Figure 3.9) for the study of bovine serum albumin (Ashby, 1969). 0·05 per cent BSA (Sigma) was dissolved in a buffer composed of 0·15 M NaCl, 0·02 M sodium acetate and 0·03 M acetic acid at pH 4·4. D_2O was used for one experiment and water for another in order to use the method of Edelstein and Schachman (1967) to obtain the partial specific volume as well as the molecular weight. The protein was dialysed for 24 hours before the runs against the buffer. The solvent channel contained 100 μl of fluorochemical FC-43 to make a lower layer and 80 μl of solvent while the solution channel contained 110 μl of FC-43 and 60 μl of solution; these quantities gave the solvent a slight overlap at each end of the solution column. Three black and two white fringes were measured and a mean value obtained. The rotor temperature was 19·4°C

* In practice the M_w cannot usually be determined because of lack of resolution of the fringes near the bottom of the cell.

and the rotor speed was 27,640 rev/min; the H_2O solvent had a density of 1·0235 and the D_2O solvent a density of 1·1075. Photographs were taken after 23 hours through sapphire windows, and a correction for distortion was applied from a control experiment. Figure 6.4 shows the plots of $\log y$ against x^2 obtained considering each fringe to be an x–y plot.

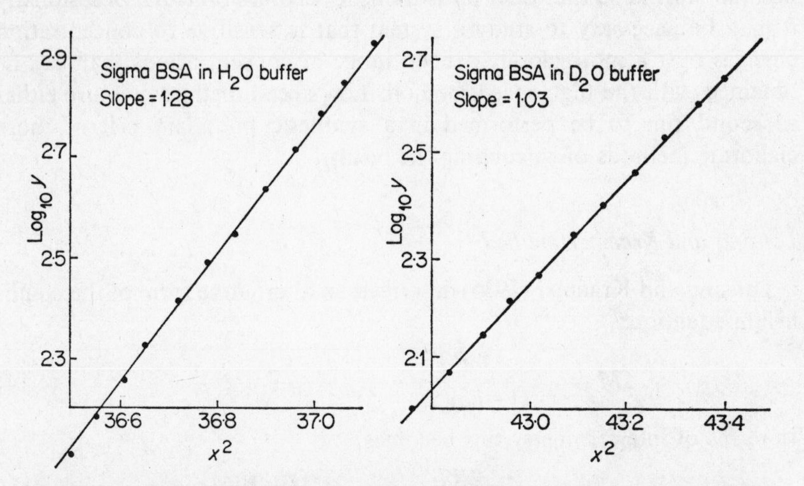

Figure 6.4. Data from a class experiment (Ashby, 1969) for Sigma BSA using a multichannel centrepiece and a Beckman Model E analytical ultracentrifuge. $\text{Log}_{10}\, y$ is plotted against x^2 where y is the vertical displacement of the fringe and x is the radial distance

Using Edelstein and Schachman's equations (5.18) and (5.19), for H_2O:

$$M(1-\bar{v}\rho) = \frac{2RT}{\omega^2} \cdot \frac{\mathrm{d}\ln c}{\mathrm{d}x^2} \tag{6.12}$$

for D_2O:

$$kM\left(1-\frac{\bar{v}}{k}\rho\right) = \frac{2RT}{\omega^2} \cdot \frac{\mathrm{d}\ln c}{\mathrm{d}x^2} \tag{6.13}$$

Assuming a value for the effect of D_2O binding of $k = 1{\cdot}01395$ for the slightly diluted D_2O used, these equations can be solved to yield $\bar{v} = 0{\cdot}737$ and $M = 69{,}600$. The linearity of the $\log y$ against x^2 plots indicates homogeneity of the Sigma preparation and ideal behaviour of the solvent–solute system. However, such linearity of the plot is not too sensitive and there may be hidden impurities.

Low speed equilibrium methods

The meniscus depletion method described above suffers from the disadvantage that, for lower molecular weights, speeds of rotation are required that may be excessive and produce problems of cell window distortion even when using sapphire windows. Interference fringes may become blurred so that their measurement becomes difficult. Occasionally it may be necessary to study a system that is sensitive to concentration changes over a more closely defined range of concentrations than can be obtained with the high speed method. Low speed methods require either a second run to be performed in a synthetic boundary cell or more elaborate methods of calculating the results.

Lansing and Kraemer method

Lansing and Kraemer (1935) described an alternative form of the equilibrium equation:

$$M = \frac{2RT}{\omega^2(1-\bar{v}\rho)(x_b^2-x_m^2)} \cdot \frac{c_b-c_m}{c^0} \tag{6.14}$$

In terms of interferometry this becomes:

$$M = \frac{2RT}{\omega^2(1-\bar{v}\rho)(x_b^2-x_m^2)} \cdot \frac{\delta j \text{ (eq. run)}}{\delta j \text{ (SB run)}} \tag{6.15}$$

The low speed equilibrium run produces a fringe pattern as shown in Figure 6.2(c). Reading the plate from left to right the number of fringes crossed between x_m and x_b (from X to Y) on a horizontal axis corresponds to δj (equilibrium run). A second run is carried out using a synthetic boundary cell (Figure 3.9) to layer solvent over the same solution to produce a fringe pattern as shown in Figure 6.2(a). A count of fringes and part fringes, crossed from solvent to solution across the boundary is δj (SB run). Application of Equation (6.15) then produces a weight-average molecular weight.

This method can give rise to considerable errors unless precautions are taken. The equilibrium experiment has been of considerable duration and at a higher speed than is usually required to form a boundary in the synthetic boundary run. Aggregated material may have spun out in the first run but it can contribute to the fringe count in the SB run. The method can be particularly unsatisfactory where there may be large amounts of a solute such as urea in the solvent system. Errors can arise due to evaporation or absorption on the cell walls so this method is best used in conjunction with other confirmatory evidence.

LaBar's method

LaBar (1965) described a method in which an equilibrium run at low speed (no meniscus depletion) is followed by a short period at high speed at the end of the run to deplete the meniscus. A problem with the low speed method is to calibrate the vertical concentration axis of the Rayleigh plots of concentration (actually refractive index) against radial distance, since there is no depleted region of zero solute concentration near the meniscus. LaBar showed that the downward vertical displacement of the fringes at the meniscus when the rotor was run faster could provide the missing information, either by direct observation or by calculation. At first sight it seems that the Yphantis (1964) meniscus depletion method could have been used in any case but the justification for LaBar's approach is that the Yphantis method cannot provide data very near the cell bottom where fringes are rising very sharply, and hence the M_w obtained does apply to the whole column of liquid in the cell. Data are also more reliable for associating systems. The high speed part of the run is of short duration only and critical sharpness of fringes is of less importance so long as the fringe displacement can be measured.

Direct observation of the downward displacement of the fringe ('relaxation') near the meniscus as the rotor is taken to the higher speed is helped by the use of a swinging slit, as in the Beckman apparatus, which enables a large number of pictures of a narrow region in the cell to be photographed on a plate.

The calculation of the fringe relaxation with time follows reasoning starting with Equation (6.16) (compare Equation (6.2)):

$$\frac{\mathrm{d} \ln j_x}{\mathrm{d}x^2} = \frac{(1-\bar{v}\rho)\omega^2}{2RT} \,.\, M_w \tag{6.16}$$

For simplicity we can write this using A to replace part of the equation. Thus:

$$\frac{\mathrm{d} \ln j_x}{\mathrm{d}x^2} = A \,.\, M_w \tag{6.17}$$

Integration yields:

$$j_x = j_m \exp \int_{x_m}^{x} AM_w \,\mathrm{d}x^2 \tag{6.18}$$

where j_m is the fringe number and m refers to the radial position of the meniscus. Differentiating the last equation and taking the natural logarithm of the result yields:

$$\ln \frac{\mathrm{d}j_x}{\mathrm{d}x^2} = \ln j_m \,.\, A \,.\, M_w + \int_{x_m}^{x} AM_w \,\mathrm{d}x^2 \tag{6.19}$$

Expanding (6.19) as a Taylor's series about x_m^2 we have:

$$\ln \frac{\mathrm{d}j_x}{\mathrm{d}x^2} = \ln j_m \,.\, AM_w + AM_w(x^2 - x_m^2) + 0(x^2 - x_m^2)^2 \tag{6.20}$$

Hence, a plot of $\ln \mathrm{d}j_x/\mathrm{d}x^2$ against $x^2 - x_m^2$ has an intercept of $\ln j_m \,.\, AM_w$ and a limiting slope at the meniscus of AM_w; the ratio (antilog intercept/slope) $= j_m$.

In practice the parameters of Equation (6.20) are obtained by using finite differences instead of differentials, but the result enables one to identify the fringe. LaBar's paper should be read for fuller details and also alternative methods of calculation leading to various types of averages.

This method was used by Ashby (1969) on a commercial sample (Koch–Light Ltd) of ribonuclease A quoted in the catalogue as chromatographically homogeneous but its purity was not further checked. The buffer used contained 0·1 M NaCl and 0·01 M NaH_2PO_4 adjusted to pH 6·4 with NaOH. The protein was dialysed against the buffer for 24 hours before use. The dialysis tubing had been boiled with EDTA and sodium carbonate to remove plasticizer and been acetylated with 10 per cent acetic anhydride in pyridine for 6 hours to reduce the pore size of the dialysis membrane. 75 μl of solution gave a height of liquid of approximately 2 mm. FC-43 fluorocarbon oil was used to give a sharp lower meniscus. Two runs were carried out, one in water solution and another in D_2O solution. Assuming a diffusion coefficient of $D^0_{20,w} = 1{\cdot}19 \times 10^{-6}$ cm^2/sec, it was calculated that 62 hours would be needed to ensure equilibrium. The rotor temperature was 19·4° and the rotor speed was 21,804 rev/min. At the end of the equilibrium run the rotor was accelerated to 47,660 rev/min for four hours to deplete the meniscus of RNase A. In water solution, $\rho = 1{\cdot}00397$ g/ml and $\bar{v}$ was taken to be 0·695 ml/g. Values of fringe displacement vertically were measured as a function of radial distance (Figure 6.4). The values of $\log \mathrm{d}j_x/\mathrm{d}x_2$ against $x^2 - x_m^2$ were plotted. The limiting slope was 0·206 and the intercept 1·17, hence from Equation (6.20):

$$j_m = \frac{\text{antilog } 1{\cdot}17}{0{\cdot}206 \times 2{\cdot}303} = 31{\cdot}20 \text{ microns}$$

This value was also determined by direct measurement from the fringe relaxation and gave an answer that proved to be identical after correcting the observed fringe relaxation at the meniscus for the fringe shift in the air–air region due to increasing the speed as described by LaBar. From the limiting slope at the meniscus $0{\cdot}206 = AM_w/2{\cdot}303$ we find that:

$$M_w = \frac{2 \times 2{\cdot}303 \times 8{\cdot}314 \times 10^7 \times 292{\cdot}6 \times 0{\cdot}206}{(1 - 0{\cdot}695 \times 1{\cdot}00397)(21804 \times 2\pi/60)^2} = 14{,}600$$

A similar calculation on the D_2O run after allowing for D_2O binding gave an answer of 13,400. The literature value from the known structure of ribonuclease A is 13,683.

Charlwood (1965, 1967) has described an approach in some ways the converse of the above. Instead of checking fringe relaxation by overspeeding to deplete the meniscus as LaBar did, a synthetic boundary cell is used to layer solvent over solution; the concentration at the meniscus then rises to the equilibrium value as the run progresses.

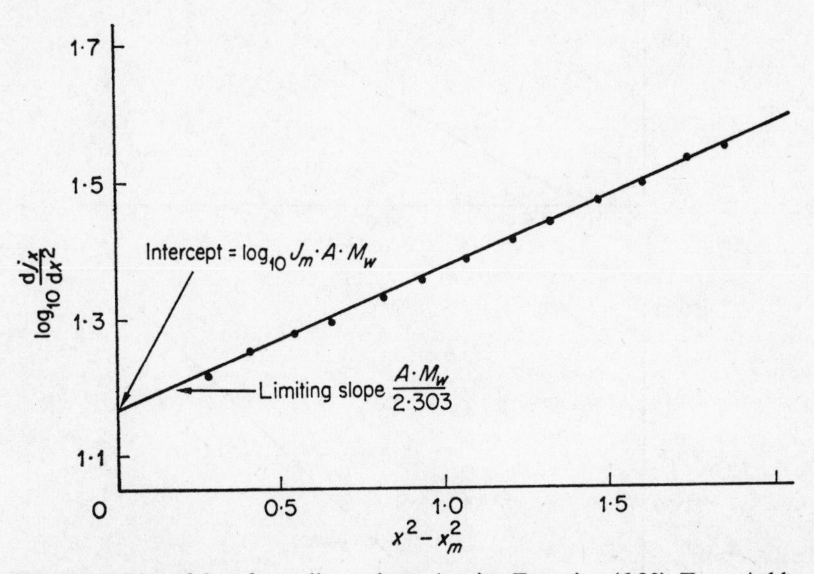

Figure 6.5. Plot of data from ribonuclease A using Equation (6.20). From Ashby, (1969)

Methods of Van Holde and Baldwin and of Lamm

These methods (Van Holde and Baldwin, 1958; Lamm 1929) produce a *z*-average result. Problems of adsorption on the cell walls and evaporation effects can give trouble where a synthetic boundary cell has to be pressed into service to obtain c^0 values. However, these methods, like some described above, work with a single equilibrium run where it has not been possible to use the meniscus depletion method. The relevant equation (Van Holde and Baldwin) is:

$$M_z = \frac{RT}{\omega^2(1-\bar{v}\rho)} \cdot \frac{\mathrm{d}}{\mathrm{d}y}\left(\frac{1}{x} \cdot \frac{\mathrm{d}y}{\mathrm{d}x}\right) \qquad (6.21)$$

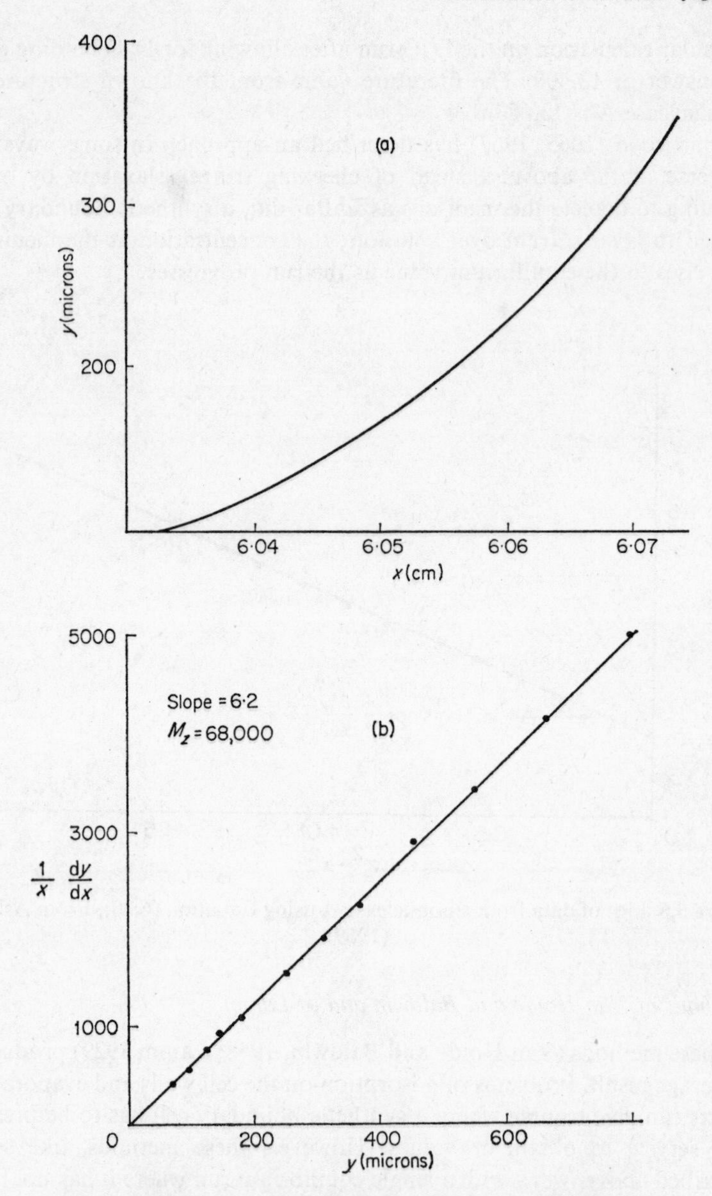

Figure 6.6. Data for Van Holde and Baldwin calculation of molecular weight of bovine serum albumin. (a) Plot of part of fringe pattern on photographic plate. (b) Data computed from (a) to obtain a molecular weight using Equation (6.21). From Ashby (1969)

Thus, if a fringe is taken to be a plot of y against x, the radial distance, it is possible to apply (6.21) using graphical methods. Ashby (1969) analysed data for bovine serum albumin (Calbiochem) in D_2O buffer under the conditions $\omega = 2896{\cdot}4$ radians/sec, $\bar{v} = 0{\cdot}734$ ml/g, $\rho = 1{\cdot}0979$ g/ml and $T = 292{\cdot}6$°K.

The fringes were measured and a plot of y against x prepared. From the slopes at increments along this, $\mathrm{d}y/\mathrm{d}x$ values were obtained. Finally, a plot of $\frac{1}{x} \cdot \frac{\mathrm{d}y}{\mathrm{d}x}$ against y was prepared from which the slope required for (6.21) was obtained. Proportionality constants between y and solute concentration disappear when plotting slopes. Precision suffers in that a second derivative is being determined but the method is simple. The data from Figure 6.5 when substituted in Equation (6.21) give:

$$M_z = \frac{8{\cdot}314 \times 10^7 \times 292{\cdot}6 \times 6{\cdot}2}{(1 - 1{\cdot}0036 \times 0{\cdot}734)(2896{\cdot}4)^2}$$

$$= 68{,}000$$

Lamm's method is described elsewhere (Chapter 10, Equation (10.3)). Although relatively little used, it is simpler to apply than Van Holde and Baldwin's method. However, due to lack of precision in the data, it is appreciably less sensitive than the latter.

The approach-to-equilibrium (Archibald) method

Since no solute can pass through the air–solution meniscus or out through the bottom of the cell, an equation similar to the equilibrium equation may be used (Equation (6.3)). In this transient state method (Archibald, 1947) the ratio of s/D is being determined from data at the upper and lower limits of the liquid column (Equation (5.10)):

$$\omega^2 \frac{s}{D} = \frac{(\mathrm{d}c/\mathrm{d}x)_m}{x_m c_m} = \frac{(\mathrm{d}c/\mathrm{d}x)_b}{x_b c_b} \qquad (6.22)$$

In practice the data from the cell bottom are not used, although in theory the same answer should be obtained as from the data at the upper meniscus, x_m. At the cell bottom there are high pressures operating and any aggregated material will affect the result. The main problem is, therefore, to determine concentration and the gradient of concentration at the upper meniscus. This is by no means simple since extrapolations have to be made to the meniscus. The focusing of the optical system is of great importance; the midpoint of the meniscus is usually taken to be its true position but at low speeds it might be more correct to take a position one-third of the meniscus thickness from the solution side. The review by Creeth and Pain (1967) draws attention to these and many other points. The advantage of the Archibald method is the short duration of the run, since a peak is only partly pulled away from the meniscus (Figure 6.7) so that even proteins of low molecular weight can be studied without making excessive demands on time or speed of rotation. The disadvantages are the fairly tedious computation of the answer and the requirement for a second run in a synthetic boundary cell with its associated errors. A very large macromolecule may require rotor speeds so low that wobbling may give trouble. The short duration of the run is useful when there is a reversible aggregation of the material that is fairly slow in reaching equilibrium; a longer duration of equilibrium run would obviously favour production of the reaction product. In this elementary account an example is given below, but anyone wishing to do research using the method must read further since it has numerous pitfalls both in theory and in practice. Nowadays equilibrium methods described above are much to be preferred except where short duration runs are required for the study of chemically interacting systems. Non-ideality and heterogeneity may cancel each other's effect, as in the case of the equilibrium method. If a θ solvent can be found then the effect of polydispersity may be revealed.

A weight average result is obtained but the method of Trautman (1956) can also give an approximate z-average answer where the solutes have similar frictional properties. A separate determination of c^0 is not required in this method. A zero time value, M_w^0, can be found by extrapolation of the experimental M_w values (unlike the equilibrium method) and this should be of value when studying polydisperse systems that redistribute themselves. Creeth and Pain (1967) warn that extrapolations of M_w^{app} against $\sqrt{t}$ can, however, be markedly non-linear. The same authors state that the precision of the method is 2 per cent or better and is maximal when the ratio of c_{x_m}/c^0 is approximately 0·75.

The following example taken from a class experiment on albumin illustrates one way of working out the molecular weight. The temperature was 293°K, the rotor speed was 24,630 rev/min, a value for $\bar{v}$ of 0·74 ml/g was assumed and the solvent was assigned a density of unity. The phase plate setting was 50° (Beckman ultracentrifuge) for the approach to equilibrium run and also the synthetic boundary run so its value does not enter into the calculation. The enlargement factor from cell to plate was measured by photographing a graticule in the position occupied by the cell in the rotor and was assigned the value $F = 2{\cdot}00$. The value of $(\mathrm{d}c/\mathrm{d}x)_m$ was 1·389 cm as a measurement directly on the plate; this was a crude value and extrapolations (mentioned below) are necessary to determine it in careful work. The concentration at the meniscus, c_m, has to be determined indirectly from the original solute concentration, c^0, and a graphical integration as shown by Equation (6.23).

All these concentrations can be handled as area measurements on the photographic plate. The original solute concentration, c^0, can be found from the area of the synthetic boundary peak under similar setting of the optical system, providing it is not allowed to sediment before taking the Schlieren picture; if this happens a radial dilution correction would have to be applied. The plateau concentration is obtained by dividing the incipient peak (Figure 6.7) up into vertical strips of equal breadth (δx) and height (δy) so that each strip has an area of $\delta x \,.\, \delta y$ cm² on the plate.

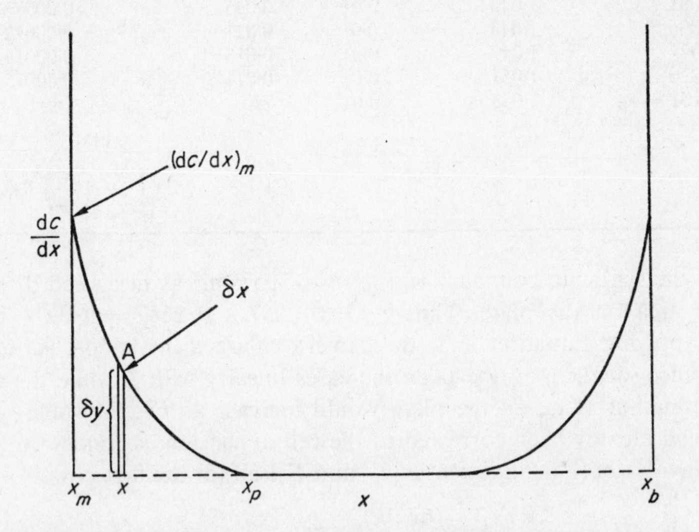

Figure 6.7. Method of measurement of Schlieren picture from Archibald run. Although concentration units are shown there is no need to convert the y coordinates to actual concentrations since the proportionality factors cancel out, in the main, in the calculation. For a detailed description see text

The areas of these strips are further corrected for radial dilution by the factor x^2/x_m^2 where x is the radial distance from the centre of rotation of each strip. Thus, Equation (6.23) is a graphical integration (a trapezoidal integration) as shown below:

$$c_m = c^0 - \sum_{x_m}^{x_p} \left(\frac{x}{x_m}\right)^2 \delta x \,.\, \delta y \qquad (6.23)$$

The data are set out in the form of a table (Table 6.1).

Table 6.1. Measurements from an approach-to-equilibrium run. For details see text

x, the radial distance expressed in cell dimensions (cm)	*Radial dilution correction* $(x/x_m)^2$	*Horizontal increment as actually read on plate,* δx (cm)	*Vertical reading,* δy, *on plate* (cm)	*Values of* $(x/x_m)^2\delta x\,.\,\delta y$ *for each element of area* (cm²)
$5{\cdot}946 = x_m$	1·000	0·01	$1{\cdot}3894 = (\mathrm{d}c/\mathrm{d}x)_m$	0·01389
5·951	1·002	0·01	1·546	0·01548
5·956	1·003	0·01	1·633	0·01638
5·961	1·005	0·01	1·686	0·01694
5·966	1·007	0·01	1·725	0·01736
etc	etc	etc	etc	etc
6·071	1·042	0·01	0·086	0·00090
6·076	1·044	0·01	0·072	0·00075
6·081	1·046	0·01	0·045	0·00047
6·086	1·048	0·01	0·021	0·00022
6·091	1·049	0·01	0·013	0·00014
6·096	1·051	0·01	0·014	0·00015
$6{\cdot}101 = x_p$	1·053	0·01	zero	zero
				Total: $\left[\sum_{x_m}^{x_p}\left(\frac{x}{x_m}\right)^2\delta x\,.\,\delta y\right]$

From the synthetic boundary run, $c^0 = 0{\cdot}3657\ \mathrm{cm}^2$ as measured directly as an area on the plate. Thus $c_m = 0{\cdot}3657 - 0{\cdot}2682 = 0{\cdot}0974\ \mathrm{cm}^2$.

In applying Equation (6.3) the camera enlargement factor has to be used since $(\mathrm{d}c/\mathrm{d}x)_m$ on the plate increases linearly with F while the area determination of c_m on the plate would increase as F^2; the value of x_m used had already been corrected to the cell dimension, so Equation (6.3) must involve both the quotient of F^2 and F itself (or the first power of F):

$$M = \frac{RT}{(1-\bar{v}\rho)\omega^2}\cdot\frac{(\mathrm{d}c/\mathrm{d}x)_m}{x_m c_m}\,.\,F \tag{6.24}$$

$$= \frac{8{\cdot}313\times10^7\times293\times1{\cdot}3894\times2{\cdot}00}{(1-0{\cdot}74\times1)(24630\times2\pi/60)^2\times5{\cdot}9462\times0{\cdot}0974}$$

$$= 65{,}900$$

The Archibald method tends to give lower answers than equilibrium methods since higher solute concentrations are used.

In the above example the Archibald method has only been described in outline. Many refinements have been suggested in the literature. Either interference or Schlieren optics may be used, but the latter is, on balance, more convenient providing that the optics are carefully aligned and the light source slit is at the focal plane (Trautman, 1958). Ehrenberg (1957) has described a technique that involves pulling the Schlieren peak further out from the meniscus to enable the top to strike the meniscus at right angles; although this would facilitate the measurement of the intercept it is not easy in practice to be sure that the right moment is chosen. Ginsburg and coworkers (1956), and Richards and Schachman (1959) have described methods of extrapolating the dc/dx plot to the meniscus. LaBar (1966) has studied the conditions under which a linear extrapolation is justified; he concluded that a linear extrapolation is justified; he concluded that a linear extrapolation to the meniscus was justified if the product of the diffusion coefficient, D, and time t, were of the order of 10^{-2} cm^2 and ε (where $\varepsilon = 2D/x_m^2\omega^2 s$) was greater than 0·5. A major difficulty has been to estimate the true position of the meniscus accurately. The method gives the molecular weight at the meniscus and this must be remembered when interpreting results. For example, Mueller (1964) performed experiments on myosin at different speeds of rotation and obtained an answer of 524,000; previous discrepancies in Archibald estimates of its molecular weight were due to concentration dependance.

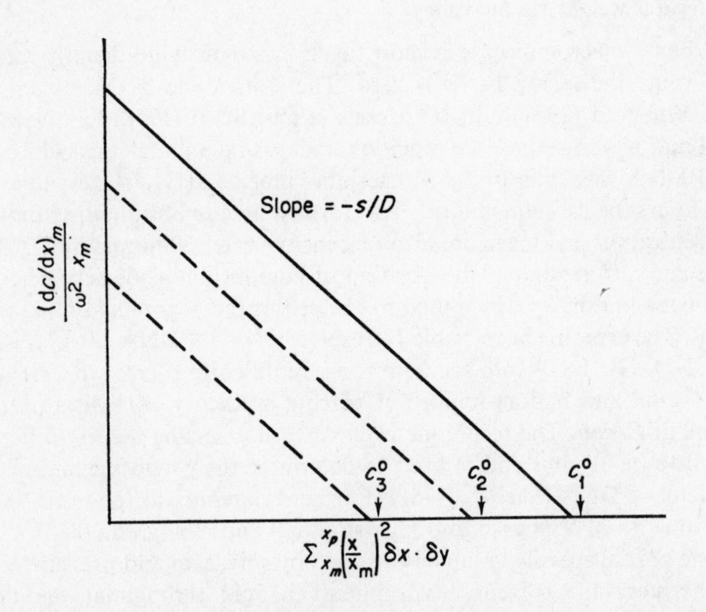

Figure 6.8. Trautman plot conveniently summarizes results of several Archibald runs on one graph

Trautman (1956) has described a convenient method for the handling of data from an Archibald series of runs. Equation (6.22) can be written in the alternative form:

$$\frac{\left(\dfrac{\mathrm{d}c}{\mathrm{d}x}\right)_m}{\omega^2 x_m} = \frac{s}{D}c^0 - \frac{s}{D}\sum_{x_m}^{x_p}\left(\frac{x}{x_m}\right)^2 \delta x \, . \, \delta y \qquad (6.25)$$

This equation can be plotted as a straight line (Figure 6.8) and is a useful way of summarizing the data from several runs. Runs from different concentrations form a series of parallel lines which are sometimes 'normalized' to one concentration and made to lie on each other by moving the lines sideways. A weight average result is obtained but it is also possible to obtain an approximate *z*-average result from the method where the solute molecules have similar frictional properties (Yphantis, 1959).

Molecular weight via buoyancy

When a macromolecule is centrifuged in a density gradient it comes to rest when the factor $1-\bar{v}\rho$ is zero. The work done by Meselson, Stahl and Vinograd (1957) and by Meselson and Stahl (1958) has blazed the trail and has since become much quoted. Using an analytical ultracentrifuge DNA was spun in 7·7 M caesium chloride at 31,410 rev/min. After five hours the caesium chloride had formed an equilibrium distribution of concentration and therefore also of density. After 30 hours the DNA had collected into a zone at the position of equilibrium isodensity whereas it had been uniformly distributed in the cell at the beginning of the experiment. The experiment was able to resolve ^{14}N–^{14}N-DNA ^{15}N–^{15}N-DNA and N^{14}–N^{15} DNA into separate zones representing density differences of 0·007 g/ml and had an important bearing on theory of replication of the DNA in *E. coli.* The technique is capable of detecting density differences down to 0·001 g/ml and is used to determine the guanosine and cytosine content of DNA (Sueoka, 1961). Further references to the technique are in the work of Vinograd and Hearst (1962) and Vinograd (1963).

The partial specific volume is affected by solvation and pressure effects; furthermore the solvent environment changes throughout the density gradient. If $\bar{v}_0$ refers to the partial specific volume under the experimental conditions, then the molecular weight, M_0, under those experimental conditions is given by the equation:

$$M_0 = \frac{RT}{\sigma^2 \bar{v}_0 (\mathrm{d}\rho/\mathrm{d}x)\omega^2 x_0} \tag{6.26}$$

where σ is the standard deviation of the distribution of solute in the band, $\mathrm{d}\rho/\mathrm{d}x$ is the effective density gradient made up of the diffusible solute, compression effects and the solvated polymer, and x_0 is the radial distance of the centre of the band. If $\delta = x - x_0$ an equation for the mass distribution in the band can be given:

$$\frac{m}{m_0} = \exp\left(\frac{-\delta^2}{2\sigma^2}\right) \tag{6.27}$$

Data required for caesium chloride are given by Thomas and Berns (1961), and for caesium sulphate, which allows steeper gradients, by Ludlum and Warner (1965). From Equation (6.27) a log plot of m/m_0 against δ^2 is linear and enables σ^2 to be obtained for substitution in Equation (6.26). Absorption optics are particularly suitable for the measurement of the concentrations. The form of the plot obtained can give valuable information on heterogeneity of molecular weight or solute density (Figure 6.9).

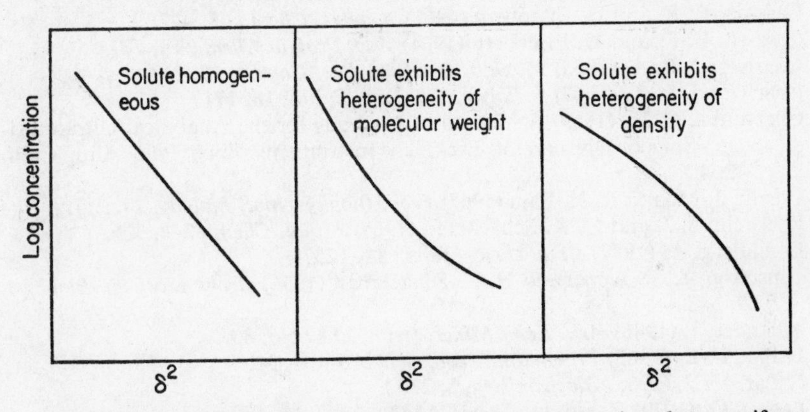

Figure 6.9. Behaviour shown during equilibrium isodensity ultracentrifugation subject to homogeneity of solute or heterogeneity in molecular weight or density

Casassa and Eisenberg (1961, 1964) have described a different approach that yields the molecular weight in unsolvated form. Data of great accuracy are needed, however, in order to apply it.

When there is heterogeneity of density it may produce considerable inaccuracy in the determination of molecular weight. This has been discussed at length in the review by Creeth and Pain (1967). It is important, therefore, to be able to recognize heterogeneity of solute density when it occurs. Schumaker and Wagnild (1965) have described a technique in which the temperature is changed during a run. If the sample has a weight distribution the different components migrate to the new isodensity position at different rates.

The equilibrium isodensity method is useful for samples that are known to be uniform in density, for example some DNA preparations. Large molecules are more convenient to study than small because they band over a shorter distance where the density gradient is more nearly linear. The theoretical background is good and the measurements are relatively simple.

REFERENCES

Archibald, W. J. (1947) *J. phys. Colloid Chem.*, **51,** 1204.

Ashby, P. (1969). Final Year Project, Department of Biochemistry, University of Leeds, England.

Aten, J. B. T. and J. A. Cohen (1965) *J. mol. Biol.*, **12,** 537.

Casassa, E. F. and H. Eisenberg (1961). *J. phys. Chem.*, **65,** 427.

Casassa, E. F. and H. Eisenberg (1964) *Adv. Protein Chem.*, **19,** 287.

Charlwood, P. A. (1965) *Biophys. Biochem. Res. Comm.*, **19,** 243.

Charlwood, P. A. (1967) *J. Polym. Sci. Part C*, No. **16,** 1717.

Chervenka, C. H. (1969) A Manual of Methods for the Analytical Ultracentrifuge, Spinco Division of Beckman Instruments, Inc., Palo Alto, Calif. 94304.

Creeth, J. M. and R. H. Pain (1967) *Prog. Biophys. mol. Biology*, **17,** 217.

Edelstein, S. J. and H. K. Schachman (1967) *J. biol. Chem.*, **242,** 306.

Ehrenberg, A. (1957) *Acta. chem. scand.*, **11,** 1257.

Ginsburg, A., P. Appel and H. K. Schachman (1956) *Archs Biochem. Biophys.*, **65,** 545.

Jullander, I. (1946) *Ark. Kemi Miner. Geol.*, **21A,** No. 8.

LaBar, F. E. (1965) *Proc. natn. Acad. Sci. U.S* , **54,** 31.

LaBar, F. E. (1966) *Biochemistry*, **5,** 2362.

Lamm, O. (1929) *Z. physik. Chem.*, A**143,** 177.

Lansing, W. D. and E. O. Kraemer (1935) *J. Am. chem. Soc.*, **57,** 1369.

Ludlum, D. B. and R. C. Warner (1965) *J. biol. Chem.*, **240,** 2961.

Meselson, M., F. W. Stahl and J. Vinograd (1957) *Proc. natl. Acad. Sci. U.S.*, **43,** 581.

Mueller, H. (1964) *J. biol. Chem.*, **239,** 797.

Richards, E. G. and H. K. Schachman (1959) *J. chem. Phys.*, **63,** 1578.

Schumaker, V. N. and J. Wagnild (1965) *Biophys. J.*, **5,** 947.

Sueoka, N. (1961) *J. mol. Biol.*, **3,** 31.

Thomas, C. I. and K. I. Berns (1961) *J. mol. Biol.*, **3,** 277.

Trautman, R. (1956) *J. phys. Chem.*, **60,** 1211.

Trautman, R. (1958) *Biochim. Biophys. Acta*, **28,** 417.

Van Holde, K. E. (1967) *Fractions, No. 1*, Beckman Instruments Inc., Palo Alto, Calif., U.S.A.
Van Holde, K. E. and R. L. Baldwin (1958) *J. phys. Chem.*, **62,** 734.
Vinograd, J. (1963) *Meth. Enzymol.*, **6,** 854.
Vinograd, J. and J. E. Hearst (1962) *Fortschr. Chem. org. Natstoffe*, **20,** 372.
Yphantis, D. A. (1959) *J. phys. Chem.*, **63,** 1742.
Yphantis, D. A. (1960) *Ann. N.Y. Acad. Sci.*, **88,** 586.
Yphantis, D. A. (1964) *Biochemistry*, **3,** 297.

CHAPTER 7

The determination of molecular conformation*

In Chapter 1, a proportionality between the force applied to a particle and the velocity produced was established. The link between these quantities was the frictional coefficient (i.e. Force $= f \times$ velocity). Various mathematical treatments are available to relate the frictional coefficient to the shape and size of the particle. Stokes' equation (Equation (1.2)) has already been given for the simple case of a spherical particle. Equation (1.9) expresses the frictional coefficient in terms of quantities determined in the analytical ultracentrifuge; a study of this equation shows that sedimentation is related to both the particle mass (and hence its volume) and the frictional coefficient, which is in turn dependent on particle shape. There is an approximate correlation between the size of the frictional coefficient and the influences of shape and mass on sedimentation behaviour, although the theoretical background to such correlations is weak and results are only roughly quantitative. In particular the effects of solvent on the particle are introduced using models of doubtful validity.

A more satisfactory approach is to use data obtained by a different technique in conjunction with information from the ultracentrifuge. In particular, viscosity measurements are particularly valuable. These can be related to size and shape of particles in the same way as ultracentrifuge data. Therefore, two equations become available to solve the two unknowns of size and shape. The theoretical advantages are considerable because no *ad hoc* assumptions have to be made about the effects of hydration; the size and shape refer to the molecule as it exists in solution.

Unfortunately, although the use of viscosity data appears to have great advantages over the earlier frictional approach, both methods suffer from the necessity of assuming models of known shapes which may not closely agree with the conformation of the dissolved macromolecule. Favourite shapes postulated include long thin rods, prolate ellipsoids (rugger ball

* See page 146 for further information.

shape), oblate ellipsoids (bun shape), randomly kinked coils and so on. There are, of course, many approaches to conformation other than by using viscosity and sedimentation data, but they will not be dealt with here. The review by Jen Tsi Yang on the viscosity of macromolecules in relation to molecular conformation can be referred to for further details (Jen Tsi Yang, 1961). Charles Tanford (Tanford, 1968) has reviewed the subject of protein denaturation and he includes a survey of the methods available for the study of conformation. Both reviews are invaluable as supplements to the present treatment.

Sedimentation data and conformation

The equations

The use of the relationship between force and velocity (Equation (1.1)) leads to Equation (7.1) (i.e. the same as Equation (1.9)):

$$\frac{M}{N}(1-\bar{v}\rho) = fs \tag{7.1}$$

Hence, knowing M, the value of the sedimentation coefficient leads to the frictional coefficient. The information obtained from f is the same as that obtained from measurements of the intrinsic viscosity. However, as Tanford (1968) points out, s measurements, although more convenient to carry out than viscosity measurements, yield less precise information; the frictional coefficient for spheres is proportional to the radius (Equation (1.2)) whereas the intrinsic viscosity is related to volume of the particle. Equation (7.1) reveals another disadvantage in that the partial specific volume is required and this may be affected by specific interactions with the components of the solvent system, whereas the intrinsic viscosity is related directly to the hydrodynamic volume and requires no additional measurements. The frictional coefficient, f, is obtained from sedimentation data using Equation (7.1) but it may also be obtained from the diffusion coefficient using the following equation (i.e. Equation (4.11)):

$$D = \frac{RT}{Nf} \tag{7.2}$$

It is customary to refer to the frictional ratio, f/f_0, where f_0 refers to the frictional coefficient of a sphere with the same volume as the particle in question. The value of f_0 can be found by applying Stokes' law coupled with the relation between volume and radius of a sphere, so that we have*:

$$f_0 = 6\pi\eta_0 r = 6\pi\eta_0\left(\frac{3V_e}{4\pi}\right)^{\frac{1}{3}} = 6\pi\eta_0\left(\frac{3M\bar{v}}{4\pi N}\right)^{\frac{1}{3}} \tag{7.3}$$

* Assuming no hydration at present.

where η_0 is the viscosity of the solvent (defined later), V_e is the volume of the particle and r is the radius of the equivalent sphere. Thus the data required to determine a frictional ratio are experimentally accessible.

The effect of asymmetry and hydration

Perrin (1936) derived equations relating the axial ratios of prolate and oblate ellipsoids to the measured frictional ratios, assuming that all orientations of the ellipsoids were possible under the conditions of the experiments. It is not proposed to quote Perrin's equations here* or to deal with his equations for the case of rotational diffusion. In order to make allowance for hydration (Oncley, 1941) the frictional ratio as determined was split into two parts:

$$\frac{f}{f_0} = \frac{f}{f_h} \cdot \frac{f_h}{f_0} \tag{7.4}$$

the ratio f/f_h expressed the axial ratio of ellipsoids through the Perrin equations while the factor f_h/f_0 was supposed to make allowance for the hydration layer surrounding the particle. Kraemer (1940) proposed the following equation to make allowance for the effect of hydration:

$$\frac{f_h}{f_0} = \left(1+\frac{w}{\bar{v}\rho}\right)^{\frac{1}{3}} \tag{7.5}$$

The partial specific volume refers to an unhydrated particle, the quantity, w, is the weight of solvent of density, ρ, solvating 1 gram of the pure solute so the quantity in brackets in Equation (7.5) expresses a fractional increase in volume of the particle as a result of hydration. The power of one-third is not unexpected since there is a third-power law relating volume and radius (and frictional coefficient). Oncley (1941) combined the information expressed in Perrin's equations and the effect of hydration (Equation (7.5)) as a set of contours; an intelligent guess was made at the value of the hydration, w; next the contour relating to the measured frictional ratio was selected, and finally, an axial ratio was read off for the prolate or oblate ellipsoid.

The treatment just described has been extensively criticized. Apart from the difficulty in reconciling simple models with the conformation of proteins it can be shown that, for simple salts such as magnesium sulphate, values of $\bar{v}$ that are positive, negative or zero have to be postulated according to the experimental conditions. Proteins do not show such extreme behaviour as this but it is now realized that the effect of solvent can be

* However, Perrin's equation for f/f_0 is quoted later (Equation 10.10).

more far-reaching than a mere spreading of a layer on the surface of the macromolecule. The choice of arbitrary hydration values to make $\bar{v}$ come out right is not justified. The solvent may affect conformation as well as increase the volume. Scheraga and Mandelkern (1953) have been particularly critical of this approach and have proposed an alternative treatment involving measurements of the intrinsic viscosities; this will be discussed below.

Viscosity data and conformation

The equations

The viscosity coefficient, η, is defined as follows: consider a plane of area, A, to which a shearing force, τ (dyne/cm^2) is applied. This plane moves relatively to another parallel plane of the same area so that there is a velocity gradient, $\mathrm{d}v/\mathrm{d}y$ (sec^{-1}) between them. Newton showed that the quantities are related by the following equation:

$$\tau = \eta \cdot \frac{\mathrm{d}v}{\mathrm{d}y} \cdot A \tag{7.6}$$

The viscosity coefficient therefore has the dimensions g/cm/sec and 1 g/cm/sec is a unit of 1 poise. When Equation (7.6) is valid for a liquid over a range of shearing stresses the liquid is said to be Newtonian. Non-Newtonian behaviour is often associated with highly asymmetric solutes that may orientate themselves along the stream lines at high values of $\mathrm{d}v/\mathrm{d}y$. It is assumed that there is no turbulence between the planes of area, A, and the streamlines are undistorted. However, the introduction of solute molecules may have the effect of distorting these streamlines.

The viscosity of the solution is defined by η, while η_0 is the viscosity of the pure solvent and their ratio η/η_0 has been called the relative viscosity or (as now preferred) the viscosity ratio. Some further definitions have to be given: $(\eta-\eta_0)/\eta_0 = \eta_{sp}$, where η_{sp} is the specific viscosity, and $\eta_{sp}/c = \eta_{red}$ (c is the concentration in g/ml) where η_{red} is the reduced viscosity or viscosity number. A particularly important quantity is the intrinsic viscosity, limiting viscosity number or Staudinger index:

$$[\eta] = \lim_{c\to 0} \frac{\eta-\eta_0}{\eta_0 c} = \lim_{c\to 0} \frac{\log_e \eta/\eta_0}{c} \tag{7.7}$$

Intrinsic viscosity (c in g/dl)

Limiting viscosity number (c in g/ml) or Staudinger index

In view of the confusion over the units for c it is important that these must be stated. Although the term Staudinger Index was proposed by the International Union of Pure and Applied Chemistry in 1957 the older term, Intrinsic Viscosity, still appears to be very popular due to its long usage.

It is not proposed to discuss now the actual apparatus used for the determination of viscosity. However, having determined η, it has to be converted to values of $[\eta]$ for use in studies on molecular conformation. The necessary equation, due to Huggins (1942), can be expressed as follows:

$$\frac{\eta_{sp}}{c} = [\eta]+k'[\eta]^2 \,.\, c \tag{7.8}$$

hence a graph of the left hand side of the equation against c could be used to provide the intrinsic viscosity (it is the intercept when the curve extrapolated to $c = 0$). The coefficient k' is believed to be connected with the effects of interaction between solute and solvent. An alternative method of plotting the results is due to Kraemer (1938):

$$\frac{\log_e \eta_{rel}}{c} = [\eta]-k''[\eta]^2 c \tag{7.9}$$

Mathematically, it can be shown that $k'+k'' = 0{\cdot}5$ so that by plotting both Equations, (7.8) and (7.9), it is possible to obtain a cross-check. Yang (1961) has described other equations that may be employed but recommends using Equation (7.8) whenever possible. Sometimes proteins may denature at interfaces, particularly when the viscosities are being determined in capillary flow apparatus at high dilutions. However, equations attempting to take the data to higher concentrations may not always be reliable as a basis for the necessary extrapolations to infinite dilution.

The effect of salts on viscosity for polyelectrolytes

The viscosity of polyelectrolytes, such as proteins, depends on pH and salt environment. Electrostatic repulsions or attractions can cause the macromolecule to contract or extend itself depending on the exact disposition of the ionic charges and their signs. We have already seen that values of the sedimentation coefficient decrease as the salt concentration in the environment decreases, and that the diffusion coefficient increases as salt concentration is lowered. The ionic atmosphere contracts closer to the macromolecule as the ionic strength of the solvent system rises, so that the charges on the macromolecule become more completely shielded. The reduced viscosity rises as the salt concentration is lowered. Fuoss and Strauss (1948) expressed the effect of ionic strength, μ, on the intrinsic viscosity by the following empirical equation, involving two constants A and B:

$$[\eta] = A\left(1+\frac{B}{\sqrt{\mu}}\right) \tag{7.10}$$

In this equation the constant A has the value of the intrinsic viscosity when the ionic strength is infinite. The ionic atmosphere is subjected to shearing stresses during the determination of viscosities. The distortion of the ionic atmosphere absorbs energy and contributes to the measured viscosity; this is known as the electroviscous effect. Various theoretical approaches to the problem have been proposed (Yang, 1961) but these will not be dealt with further in this book. The main point that emerges is that, at high ionic strengths, and when working at low concentrations of the polyelectrolyte, such effects become negligibly small. Furthermore, when using intrinsic viscosity data in conjunction with sedimentation data as described below the data from both techniques should have been obtained in the same solvent environment.

Conformation models

Viscosity data related to conformation

The connection between macromolecular structure and measurements of viscosity is the subject matter for much study and speculation. The object of this section is to try and quickly reach a point at which sedimentation data can be introduced simply, and therefore much worthy material will be ignored in the interest of brevity and simplicity. In general, a spherical protein will give rise to a lower intrinsic viscosity than an extended molecule of the same molecular weight. The effect of particle asymmetry is more pronounced than molecular weight, but for a series of similar polymers intrinsic viscosities do increase with the molecular weights. Once again, it can be emphasized that intrinsic viscosity (like sedimentation data) depends on size and shape of the solute molecules.

Einstein (1906) considered the effect of introducing spherical particles into a continuous medium. The particles were considered to be large compared with the solvent molecules but small compared with the dimensions of the apparatus. They were completely wet by the solvent and no turbulence was envisaged. The particles distorted the stream lines so that at low concentrations the following equation was calculated to be applicable:

$$\frac{\eta-\eta_0}{\eta_0} = \nu\Phi \tag{7.11}$$

Thus the left hand side of the equation (specific viscosity) equalled the product of $v(= 2{\cdot}5)$, the viscosity increment, and Φ which is the volume fraction (total volume of solute/total volume of solution). The value of 2·5 for v was found for infinitely dilute solutions where Φ approaches zero. Within limits, therefore, the number of particles was not important but only their total volume.

The Einstein model has served as a basis for equations designed to express viscosity measurements as a function of asymmetry. For example, one approach was to consider a long rod to be made up of many spheres joined side by side and then to modify the Einstein equation. Simha (1940) was able to produce an equation which has been particularly successful and refers to ellipsoids of resolution with all orientations equally possible (i.e. the Brownian movement was not interfered with by an excessive ratio of shear rate to rotational diffusion coefficient, $\alpha = G/\theta$). For a prolate ellipsoid, assuming that both α and Φ approach zero, then v could be written as a function of axial ratio p:

$$v = \frac{p^2}{15}(\log_e 2p - 1{\cdot}5) + \frac{p^2}{5}(\log_e 2p - 0{\cdot}5) + \frac{14}{15} \tag{7.12}$$

and for oblate ellipsoids

$$v = \frac{16q}{15} \arctan q \text{ (where } q = 1/p.) \tag{7.13}$$

The approach to conformation using sedimentation and viscosity data

The β function of Scheraga and Mandelkern

Mehl, Oncley and Simha (1940) tabulated v in terms of p and q for a wide range of asymmetries. Their work provided the basis for a most important development by Scheraga and Mandelkern (1953) in that sedimentation data could be worked in with the viscosity results. Equation (7.11) relates viscosity measurements to a 'shape factor', F, and a volume fraction Φ but with the advantage that the shape factor was already calculated as a function of axial ratio. If V_e is the volume of the hydrodynamically equivalent ellipsoid, and, concentrations, c, are expressed as g/dl it is possible to write (7.11) in the form:

$$\eta_{sp} = \frac{NV_e c \,.\, v}{100M} \tag{7.14}$$

or

$$\frac{\eta_{sp}}{c} = [\eta] = \frac{NV_e \,.\, v}{100M} \tag{7.15}$$

Hence the intrinsic viscosity is capable of expression in terms of N

(Avogadro's number), M the molecular weight, V_e a hydrodynamically equivalent ellipsoidal volume which refers to the hydrated particle and the shape factor already tabulated in terms of p and q, the axial ratios.

Scheraga and Mandelkern now sought a second equation involving two of the unknowns of (7.15), shape and volume, and found this in existing sedimentation theory. Stokes' equation, $f_0 = 6\pi\eta_0 r_h$, could be used taking r_h as the hydrated radius of a sphere. Using the relationship between radius and volume of a sphere (V_e) the Stokes' equation becomes:

$$f_0 = 6\pi\eta_0\left(\frac{3V_e}{4\pi}\right)^{\frac{1}{3}} \tag{7.16}$$

Perrin (1936) had already calculated values of axial ratio for ellipsoids of revolution as a function of frictional ratio so, by substituting the shape factor, F (a known function of p and q) in (7.16) we get:

$$f = \frac{f_0}{F} \quad \text{or} \quad f = \frac{6\pi\eta_0}{F}\left(\frac{3V_e}{4\pi}\right)^{\frac{1}{3}} \tag{7.17}$$

Since f is experimentally accessible, Equation (7.17) relates it to volume and shape. From (7.15) and (7.17) therefore, a solution for these two unknowns can be found. The value of f can be taken from Equation (7.1) or, using diffusion measurements, via the relationship $D = RT/Nf$. If the sedimentation data are chosen, the simultaneous equations can be solved to yield:

$$\frac{(4\pi)^{\frac{1}{3}}v^{\frac{1}{3}}FN^{\frac{1}{3}}}{6\pi(300)^{\frac{1}{3}}} = \beta = \frac{Ns^0\eta_0[\eta]^{\frac{1}{3}}}{M^{\frac{2}{3}}(1-\bar{v}\rho)} \tag{7.18}$$

Thus the new 'shape factor', β, is compiled from the other shape factors v and F. This treatment opened up a new approach to the study of molecular conformation in solution that did not involve arbitrary assumptions about hydration although the simple geometry of ellipsoids of revolution was still assumed. Some values of β are given in Table 7.1.

Although the theory is more satisfactory than the earlier approach to conformation described at the beginning of this chapter, since the latter involves arbitrary assumptions about hydration, the determination of conformation continues to be a difficult problem. An early application of the theory was made, by Johnson and Rowe (1961), to myosin. They obtained the following data for the protein: $s^0_{20} = 6{\cdot}43$ Svedbergs, $D^0_{20} = 1{\cdot}05 \times 10^{-7}$ cm^2/sec (a literature value), $[\eta] = 2{\cdot}35 \pm 0{\cdot}03$ dl/g and $\bar{v} = 0{\cdot}725$, leading to a value for β of $2{\cdot}83 \times 10^6$. A value for the axial ratio of 34·5 was obtained and a value of $V_e = 2{\cdot}52$ ml/g.* They mentioned

* Recalculated from the volume of the protein particle V_e as used in the Scheraga-Mandelkern treatment.

the inconsistent and low values for the Archibald molecular weights in their paper. The value of V_e was exceptionally large and difficult to explain and would require excessive solvation to account for it. Alternatively, the volume might be considered to be a swept or excluded volume, or solvent could be trapped within the molecule. Myosin has always been a difficult and highly asymmetrical protein to study. However, apparent discrepancies from the theory can sometimes shed new light. Legumin was quoted as showing the appearance of a hollow cylinder under the electron microscope. Obviously, the theory should help in shedding light on such problems.

Table 7.1. Values of β as a function of axial ratio for ellipsoids of revolution (Scheraga and Mandelkern, 1953)

Prolate		*Oblate*	
axial ratio	$\beta(\times 10^6)$	*axial ratio*	$\beta(\times 10^6)$
1	2·12	1	2·12
2	2·13	2	2·12
3	2·16	3	2·13
4	2·20	4	2·13
5	2·23	5	2·13
10	2·41	10	2·14
15	2·54	15	2·14
20	2·64	20	2·15
30	2·78	30	2·15
50	2·97	50	2·15
100	3·22	100	2·15
200	3·48	200	2·15
300	3·60	300	2·15

If β is found to be greater than $2{\cdot}15 \times 10^6$, the oblate case is ruled out.

The theory was applied by Brown and Netschey (1967) to membrane proteins and lipoproteins, some measurements being performed in 8 M urea. They concluded that in normal solutions the molecules were expanded and highly solvated. There was no evidence that lipid affected the state of aggregation of protein in 8 M urea or greatly affected the conformation of the proteins in that solvent.

Another example of an extremely prolate ellipsoid is the stiff rod of the DNA molecule. Edwards and Shooter (1965) showed that for double-stranded native molecules behaving as rods, $\beta = 3{\cdot}1\text{–}3{\cdot}4 \times 10^6$, whereas for single-stranded, denatured random-coil molecules $\beta = 2{\cdot}5 \times 10^6$.

Conformation and the Wales–van Holde ratio

The theory of Scheraga and Mandelkern suffers from the fact that the slow change of the β function with axial ratio requires high precision in the determination of the various parameters required for its computation. A further complication arises in that β has a value for random coils similar to the value for a rod of axial ratio 15. Creeth and Knight (1965) have listed many of these considerations. V_e must not be confused with the partial specific volume $\bar{v}$ but, at least, ratios of $V_e/\bar{v}$ approaching unity indicate compactness of form (Schachman, 1959). Wales and van Holde (1954) considered the connection between conformation and the ratio $k_s/[\eta]$, in which k_s expresses the concentration dependence of the sedimentation coefficient as a reciprocal law, $1/s = 1/s^0(1+k_s c)$. This ratio has been carefully examined and the data published (Creeth and Knight, 1965) for a wide range of proteins and some organic polymers. Strict criteria were imposed on the accuracy of the data selected and interacting systems were excluded. They concluded that globular or spherical proteins give fairly high $k_s/[\eta]$ values, approaching that for model spheres, and asymmetric molecules give low values. Denatured proteins give high values similar to the values for spheres. They state that it should be possible to decide whether a compact globular structure, a random coil, or an asymmetric rigid structure is indicated. If $k_s/[\eta]$ is found to be 1·5–1·7 the molecule must be nearly spherical, but the value of the viscosity increment, ν, will then show in its departure from the value of 2·5 whether the configuration is compact or expanded. Values of $k_s/[\eta]$ lower than 1·5 indicate asymmetry.

Creeth and Knight (1967) have applied these principles to a study of the blood-group substances. Their paper is rich in experimental detail and discussion, and contains far more information than can be given in this introductory account. The s values were strongly concentration dependent and followed the reciprocal law. The limiting values were slightly temperature dependent. The intrinsic viscosities were more temperature dependent and indicated a very asymmetric or very expanded molecular conformation for the glycoproteins. The ratio of $k_s/[\eta]$ indicated a roughly spherical molecular conformation, while the Einstein viscosity equation indicated an expansion factor of about 60; the latter is compatible with a flexible configuration approaching that of a random coil. The effect of sodium dodecyl sulphate did not produce any significant change in the secondary structure or the serological activity. All the properties observed indicated the absence of any secondary structure in blood-group substances.

Tanford (1967, 1968) and his school have made a particular study of protein denaturation employing a wide range of physical techniques. The random coil configuration would depend on the choice of a solvent that would provide stronger forces of attraction with the polymer segments than exist between different polymer segments. Such a solvent would be hard to find in view of the presence of hydrophilic and hydrophobic regions within the one protein molecule. Their results indicate that concentrated aqueous solutions of guanidine hydrochloride produce generally larger changes in physical and chemical properties than occur in other denaturing media, although urea may be equally effective for many proteins. A moderate concentration of β-mercaptoethanol was introduced to disrupt disulphide bonds and to prevent their reoxidation. For linear random coils (Flory, 1953; Tanford, 1961) the following equation was used:

$$\frac{M}{Ns^0}(1-\psi'\rho) = P\eta_0\langle L^2\rangle^{\frac{1}{2}} \tag{7.19}$$

In this equation, P is a universal constant, ρ and η_0 refer to the solvent, and $\langle L^2\rangle^{\frac{1}{2}}$ refers to the end-to-end distance of the chain. The term $(1-\psi'\rho)$ was a corrected value for the usual term $(1-\bar{v}\rho)$ defined by the equation:

$$(1-\psi'\rho) = (1-\bar{v}_2\rho)+(1-\bar{v}_3\rho)\frac{\delta g_3}{\delta g_{2_{n_3}}} \tag{7.20}$$

Equation (7.20) refers to a three-component system with the subscripts 2 for the protein and 3 for the second solvent component, while the partial derivative expresses the preferential binding of component 3 in grams per gram of protein. The component ψ' can be measured directly after a long period of dialysis (Casassa and Eisenberg, 1961, 1964) and Tanford (1968) lists sources of information on the determination of the partial derivative.

Determination of intrinsic viscosity

Much could be written on the subject of viscometry. In the space of this book, however, only an outline will be given of the technique at present being applied in the protein field. Further information is available in the writings of Schachman (1957) and Yang (1961).

The coefficient of viscosity is largely dependent on the temperature. For water, the viscosity changes 2 per cent per degree at 20°C, so an accurately controlled bath is required with control to ±0·005°C or better.

Capillary viscometers are commonly used. Creeth and Knight (1967) used a viscometer with a capillary length of 80 cm and internal diameter

0·5 mm. The capacity was 1·5 mm. The capacity was 1·5 ml and the volume of flow 0·5 ml. It was found necessary to clean the viscometer with a mixture of hydrogen peroxide and sulphuric acid after every filling in order to obtain reproducible results. All dilutions of protein used were made from a single stock solution and a micrometer syringe was used to fill the viscometer. The flow times of the solution through the capillary must be measured very accurately, usually by observing the time taken for the meniscus to pass between two marks. Schachman (1957) described the use of fine lines scratched in the blacked out portion of the viscometer, so that the passage of the meniscus can be observed as a flash of light passing the scratches; with this system an electric timer was advocated. Obviously, intrinsic viscosity determinations require the timing of both solvent, and increasingly dilute protein solutions through the same capillary, so that the differences in time become very small.

The densities of the solutions must be measured. Creeth and Knight (1967) used pyknometers of 5 ml capacity, fitted with necks made of 1 mm diameter capillary tubing with reference marks scratched on them. These were filled with a syringe and the actual volume was recorded, after equilibration, by using a cathetometer to measure the distance of the meniscus from the reference mark. The capillary diameter had been previously checked by measurement of the length of a mercury thread of known weight and density in the capillary. The bath temperature was controlled to $\pm 0{\cdot}002$°C and the brass weights used were corrected for air buoyancy.

If a capillary has a length L and diameter a, and if a liquid of viscosity coefficient η is forced through it under a pressure of P dynes/cm^2, then Poiseuille's law relating these quantities can be written:

$$V = \frac{\pi P a^4}{8\eta L} \tag{7.21}$$

where V is the volume of liquid passing per unit time expressed as ml/sec. Obviously, when comparing the times t and t_0 for equal volumes of two liquids of viscosity coefficients η and η_0, Equation 7.20 can be rewritten in the form:

$$\eta_{\text{rel}} = \frac{\eta}{\eta_0} = \frac{t}{t_0} \cdot \frac{\rho}{\rho_0} \tag{7.22}$$

Thus the left hand side of the expression is the relative viscosity if η_0 refers to the solvent and η to the solution. Also:

$$\eta_{\text{sp}} = \eta_{\text{rel}} - 1 = \frac{t\rho - t_0\rho_0}{t_0\rho_0} \tag{7.23}$$

Hence a plot of η_{sp}/c and $\log_e \eta_{rel}/c$ can be plotted as ordinates on the same graph against concentrations of solute, c, as abscissae: The two plots should converge at the $c = 0$ point to give an intercept recording the intrinsic viscosity (Equations (7.8) and (7.9)).

It must be emphasized that the above description has overlooked many possible sources of error. The kinetic energy of the liquid emerging from the capillary has resulted from part of the potential energy used to drive the liquid through; when comparing similar liquids this should be negligible. There may be anomalous effects near the capillary wall, surface tension effects, drainage errors, turbulence and so on. Newtonian flow is assumed, yet the shearing stress increases linearly from zero at the centre of the capillary to a maximum at the walls. Extrapolation of data to zero rate of shear may be required where the behaviour of the liquid is non-Newtonian (Yang, 1961).

REFERENCES

Brown, A. D. and A. Netschey (1967) *Biochem. J.*, **103,** 24.
Casassa, E. F. and H. Eisenberg (1961) *J. phys. Chem.*, **65,** 427.
Casassa, E. F. (1964) *Adv. Protein Chem.*, **19,** 287.
Creeth, J. M. and C. G. Knight (1965) *Biochim. Biophys. Acta*, **102,** 549.
Creeth, J. M. and C. G. Knight (1967) *Biochem. J.*, **105,** 1135.
Edwards, P. A. and K. V. Shooter (1965) *Q. Rev. chem. Soc.*, **19,** 369.
Einstein, A. (1906) *Annln Phys.* (4), **19,** 289.
Flory, P. J. (1953) *Principles of Polymer Chemistry*, Cornell U.P. Ithaca, New York.
Fuoss, R. M. and U. P. Strauss (1948) *J. Polymer. Sci.*, **3,** 602.
Huggins, M. L. (1942) *J. Am. chem. Soc.*, **64,** 2716.
Johnson, P. and A. J. Rowe (1961) *Biochem. J.*, **79,** 524.
Kraemer, E. O. (1938) *Ind. Engng. Chem. analyt. Edn.*, **30,** 1200.
Kraemer, E. O. (1940) in *The Ultracentrifuge*, Svedberg, T. and K. O. Pedersen, Oxford U.P., Oxford.
Mehl, J. W., J. L. Oncley and R. Simha (1940) *Science*, **92,** 132.
Oncley, J. L. (1941) *Ann. N.Y. Acad. Sci.*, **41,** 121.
Perrin, F. (1936) *J. Phys. Radium* (7), **7,** 1.
Schachman, H. K. (1957) *Methods in Enzymology*, **4,** 32.
Schachman, H. K. (1959) *Ultracentrifugation in Biochemistry*, Academic Press, New York.
Scheraga, H. A. and L. Mandelkern (1953) *J. Am. chem. Soc.*, **75,** 179.
Simha, R. (1940) *J. phys. Chem.*, **44,** 25.
Tanford, C. (1961) *Physical Chemistry of Macromolecules*, Wiley, New York.
Tanford, C. (1968) *Adv. Protein Chem.*, **23,** 121.
Tanford, C. Kazuo Kawahara and Savo Lapanji (1967) *J. Am. chem. Soc.*, **89,** 729.
Wales, M. and K. E. Van Holde (1954) *J. Polym. Sci.*, **14,** 81.
Jen Tsi Yang (1961) *Adv. Protein Chem.*, **16,** 323.

CHAPTER 8

Quantitative analysis of mixtures

The analysis of a mixture of macromolecules is by no means straightforward. However, the problem can be split up into sections as follows:

1. Instrumental factors. These are concerned with the type of optical system being used, the geometry of the cell and the inhomogeneity of the centrifugal field.

2. Properties of the system being analysed. These properties include the specific refraction increment (or absorption coefficient), the relationship between sedimentation coefficient and solute concentration and the possibility of there being dynamic equilibria between the components of the mixture.

A. Instrumental factors

Many of the instrumental factors have already been described in Chapter 3, and it is only necessary here to recapitulate. When ultraviolet absorption systems are in use the familiar Beer–Lambert law (Equation (3.2)) is applied to relate concentration to optical density. If the advantages of double beam working are not available, the problem is technically more difficult and usually involves photodensitometry studies of the plates to relate them to known amounts of solute in order to provide a calibration curve.

For refractometric work, the relationship between peak area or fringe shift has already been given in terms of the geometry of the optical system and a change in refractive index (not concentration directly) due to solute dissolved in a suitable buffer solution (Equations (3.3) and (3.4).) The refraction due to the buffer is subtracted from the solution pattern in Schlieren work by the use of a double sector cell, so that the solvent baseline is correctly positioned under the solution peak. In interference work, a double sector cell is used to compare solvent with solution in any case, so that the fringe displacement should only be a function of the solute concentration. For careful work, however, a check on distortion is obtained by running the ultracentrifuge with solvent in both channels.

It is possible to measure all the optical parameters needed to relate peak areas to refraction in the cell for use in Equation (3.3). In practice, time can be saved by using a special calibration cell (Beckman Instruments Inc) involving a wedge to simulate a known increase in solute concentration towards the bottom of the cell. This produces a straight line on the Schlieren pattern parallel to the reference base line. A rectangular area is defined on the plate and measured so that a direct measure of the total contribution of the optical parameters is obtained. Alternatively, a 1 per cent solution of sucrose gives a refraction increment of 0·00143, so that its use in a synthetic boundary cell as an under layer for water will give a direct calibration of the peak area produced.

Before peak areas can be related to solute concentration they must all be corrected for the radial dilution arising from the sector shape of the cell and the inhomogeneity of the gravitational field (Equation (5.8)). Each peak is, therefore, multiplied by the factor $(x/x_m)^2$ to correct its area to that corresponding to zero time when the peak was at the air–solution meniscus.

B. Properties of the system itself

The relationship between solute concentration and light absorption or refraction

In work involving light absorption, the wavelength of light used should correspond to the absorption peak, just as in the field of spectrophotometry. Deviations from the Beer–Lambert law cannot be discussed here but are dealt with in text books on spectroscopy. In this connection the property that depends on the system itself will be the molar extinction coefficient of the solute at the chosen wavelength (Equation (3.2)).

In refraction measurements the specific refraction increment must be known if the refractive index measurements are to be related to solute concentration. Equation (3.1) has already been given to define this constant, values for which do not vary dramatically from protein to protein. For example, in the case of bovine serum albumin the specific refraction increment is 0·00187 compared with 0·00171 for a sample of β-lipoprotein associated with a considerable amount of lipid material. The wavelength of light does affect the value slightly so that, for example, the value for bovine serum albumin changes from 0·00187 at a wavelength of 578 nm to 0·00195 at 436 nm. An increase of 20°C in temperature can produce a small change in the third significant figure (Doty and Edsall, 1951). For light-scattering studies the value of the specific refraction increment is needed with great accuracy but such precision is not usually required for ultracentrifuge work. Although measurements of refractive index may be made with high precision, difficulties usually arise in correctly evaluating the concentration of protein being used for these measurements. The relationship between refractive index and concentration is usually linear. In general, it is not usually possible to measure peak areas or fringe displacements with high accuracy. The more accurate fringe measurements are claimed to be within 0·02 of a fringe but not all workers achieve this degree of precision. A knowledge of the value of the specific refraction increment is not usually available and may not be the same for all components in a mixture. The usual approach, lacking this information, is to assume that all components have roughly the same values of this constant, so that the relative quantities would be proportional to the relative areas of the Schlieren peaks after they have been corrected for radial dilution.

It is good practice, when analysing data, to calculate whether the refraction data agree with measurements made on the solution before the actual run. Lack of correlation could be due either to loss of aggregated material at the bottom of the cell or to absorption. A statement that the pattern observed corresponds to some percentage of the material actually put in the cell is then made.

The Johnston–Ogston effect

Further complications arise in that the peak areas are affected by the nature of the relationship between sedimentation coefficient and concentration (Johnston–Ogston effect) and, when there are dynamic equilibria between the macromolecules affecting their areas, *s* values and shapes (Gilbert theory).

In 1935 it was reported (McFarlane, 1935) that the analysis for a mixture of proteins apparently changed with the total concentration. The ratio of slow to fast components, as determined from the Schlieren peaks, became greater at higher concentrations. Pedersen (1936, 1938) interpreted the phenomenon as an aggregation effect. Johnston and Ogston (1946) showed this explanation was not tenable by using a coloured protein, haemoglobin, to give a visible indication of its presence relatively to the Schlieren peaks. They showed that the phenomenon that now bears their names could be explained directly as a consequence of the fall in sedimentation coefficients with concentration increase.

Johnston and Ogston showed that if two sedimenting species are under consideration there are three s values to be taken into account. The fast component always moves in an environment of slow component and has one s value, but the slow component must be ascribed two s values depending on whether the molecules are on their own or in the presence of the fast molecules; Figure 8.1 may make this clearer. Region γ extends from the fast boundary to the cell bottom while region β lies between the slow and fast boundaries. The fast component, A, has a concentration c_A^γ while the slow component, B, has concentrations c_B^β and c_B^γ depending on whether the molecules are in the β or γ regions. The velocity of the fast boundary is v, while v_B^β and v_B^γ are the velocities of the slow (B) molecules in regions β and γ. A very important part of the argument is that v_B^γ must be less than v_B^β because it is related to the region where the total concentration of solute is higher due to the additional presence of component A. This is, of course, a direct result of the fall in s values with increasing solute concentration.

At the commencement of the run region β does not exist but it develops and widens as the run progresses. Molecules of component B left behind the fast boundary run faster and close up behind the B molecules running more slowly in the γ region. Thus the concentration c_B^β becomes greater than c_B^γ. The drop in concentration of B across the fast boundary results in a reduction in the observed size of the fast Schlieren peak; this can be considered to be the resultant of a positive peak (due to component A) with a superimposed negative peak due to component B. Obviously, the negative gradient can only be detected where it is possible to apply additional measurements such as light absorption.

Consider a lamina moving with the fast boundary. By considering the relative velocities of the A and B molecules at the faces of this lamina, it can be seen that the change of amount of slow component, B, per unit area is given by:

$$\frac{\mathrm{d}n}{\mathrm{d}t} = c_B^\gamma(v - v_B^\gamma) - c_B^\beta(v - v_B^\beta) \tag{8.1}$$

The slow component would accumulate in the lamina but for the adjustment of the B concentrations to offset the differences in velocities. The net accumulation becomes zero when $\mathrm{d}n/\mathrm{d}t$ is zero. Hence:

$$c_B^\beta = c_B^\gamma \cdot \frac{v - v_B^\gamma}{v - v_B^\beta} \tag{8.2}$$

or, in terms of the corresponding sedimentation coefficients:

$$c_B^\beta = c_B^\gamma \frac{s^{\beta\gamma} - s_B^\gamma}{s^{\beta\gamma} - s_B^\beta} \tag{8.3}$$

This equation shows that the anomaly becomes greater when the two sedimenting species are close in their sedimenting velocities. The Johnston–Ogston effect becomes negligible at high dilutions, so that a practical

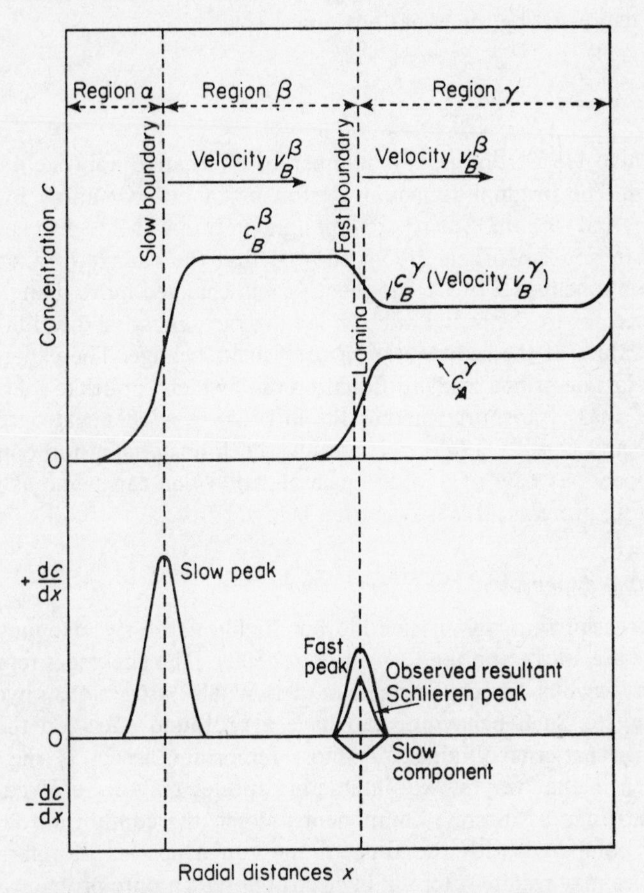

Figure 8.1. Operation of the Johnston–Ogston effect for a two component mixture. Molecules of the slow component are moving more slowly in the γ region than the β region. Accordingly, molecules on the β region pile up behind those in the γ region to produce an inverse gradient that decreases the observed size of the fast Schlieren peak

approach to the analysis of mixtures would be to extrapolate data to lower total protein concentrations. In practice this may not always be possible where solute–solute interactions occur and in any case it leads to the necessity for determining areas of peaks that are scarcely resolved from the baseline. One approach to the data required in Equation (8.3) has been to assume the same concentration dependence for the slow component when it is present singly, or together with the fast component. This is expressed by the equations:

$$s_B^\beta = s_B^0 - kc_B^\beta,$$

and

$$s_B^\gamma = s_B^0 - k(c_B^\gamma + c_A^\gamma) \tag{8.4}$$

Schachman (1959) discusses a number of alternative approaches to the problem. The original Johnston–Ogston proof did not allow for radial dilution and inhomogeneity of the gravitational field. For example, Trautman and coworkers (1954) showed that the change in area of the slow component in a two-component system changed more than could be accounted for by the radial dilution law; in fact, as radial dilution occurs, so the values of the sedimentation coefficients change. The experimental pattern just described leads to Equation (8.3) which applies to many other cases of mass transport whether the interaction is chemical or physical (Nichol and Ogston, 1965). As a result, the Johnston–Ogston expression now appears as part of a wider generalization one can make as regards interacting proteins; this is discussed below.

Interacting systems and the Gilbert theory

It has been said, in unscientific but highly expressive language, that proteins are sticky and tend to clump together. The stickiness referred to involves various forces; ionic, van der Waals' attractions, hydrogen-bonding, etc. Such behaviour can have a profound effect on the ultracentrifuge patterns obtained. A most important aspect is the kinetic approach in that very slowly interacting species may be resolved by the ultracentrifuge as discrete components. When the equilibrium is set up rapidly compared with the time of the run a species of intermediate properties may result. After all, in electrophoresis a pure protein runs as a single component although very many rapidly interconvertible ionic forms are present. However, one should mention that Cann and Goad (1968) have described conditions under which this is not necessarily so.

It has been pointed out (Green, 1969) that many protein aggregations are successive dimerizations that may proceed to fairly large aggregates. The units of the dissociation constant of a dimer are units of concentration. On this basis the lowest constants measurable by sedimentation methods are of the order of 10^{-6} M. Measurements of fluorescence or enzyme activity extend the lower limit to 10^{-9} M. In suitable hybridizing systems, measurements of sub-unit interchange may extend the limit to 10^{-13} M. Where the sub-units are bound together in a very stable assembly it may be possible to dissociate the complex using 6 M guanidinium chloride and then use the ultracentrifuge to reveal them. However, in this section consideration has to be given to a system of macromolecules in dynamic equilibrium.

Early views were that, in cases of rapid equilibration compared with the duration of the ultracentrifuge run, a system such as $nA \rightleftharpoons A_n$ would provide a single peak with properties averaged between the monomer and polymer. A system $A \rightleftharpoons B+C$ would provide two Schlieren peaks, since, assuming decreasing sedimentation velocities from A to C, the slow peak would correspond to the light molecule, C, alone whereas in the second (faster) boundary the simultaneous presence of component B would give rise to a proportion of the heaviest component A as well; however, the situation is more complicated than this. Gilbert (1955) presented a theoretical interpretation of the puzzling data presented for α-chymotrypsin by Massey, Harrington and Hartley (1955) for sedimentation behaviour at pH 7·9 and at low ionic strengths. Gilbert (1959, 1963) has given further expression to his ideas for the solution of the difficult problem of a reversible system in which a gravitational field is continually removing heavier components. He was able to interpret the results of Massey and coworkers in terms of a single peak system for dimer formation and a two peak system for the formation of hexamers. He was able to calculate the fraction of hexamer in accordance with the experimental results and was able to predict accurately the values of s as a function of concentration for both components. In his treatment the effects of diffusion had to be neglected but the effect of diffusion on the calculated peak shapes may be very great and may to some extent mask the results. Today the theory is most valuable for the interpretation of such interacting systems.

Taking the simplest case of a monomer giving a polymer, $nA \rightleftharpoons A_n$, a fruitful approach has been to consider that the polymer molecules could be considered stationary during the ultracentrifuge run while the monomer molecules, relatively speaking, flowed past them. There is a chromatographic analogy to this situation when the absorption of a component depends to a higher power than the first on concentration giving rise to a diffuse rather than a sharp leading boundary. The development of this idea leads to the prediction that in the ultracentrifuge a diffuse boundary will occur. If we let M and P represent concentrations by weight of A and A_n respectively we can write for the equilibrium the equation:

$$M^n = P \,.\, K \tag{8.5}$$

where K represents an equilibrium constant. Figure 8.2 shows the comparison with the chromatographic situation of the non-linear isotherm.

Considering Plate VIII (a), a lamina of thickness dx at a distance x from the meniscus is moving at the speed of the monomer molecules, which are considered to have zero relative velocity, while the polymer molecules have a velocity v. Monomer molecules can only be lost from the lamina by conversion into polymer molecules which, in turn, can only be lost

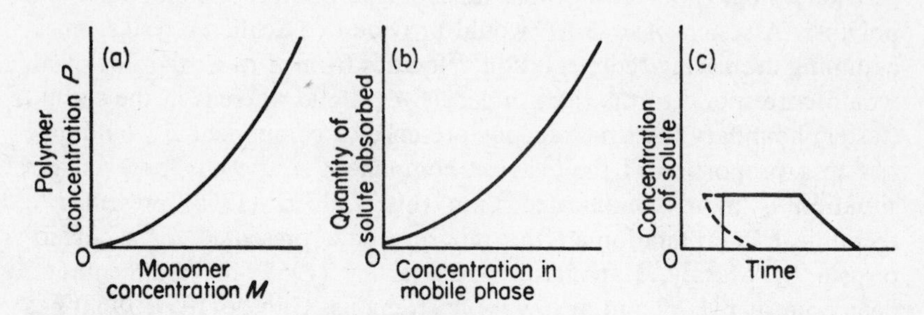

Figure 8.2. (a) Graph showing relationship between M and P according to Equation (8.5). (b) Non-linear chromatographic adsorption isotherm. (c) Non-linear isotherm in (b) gives rise to elution of a plug of solute with a sharp and a diffuse edge

from the lamina when they have different concentrations at each face of the lamina. The increase in grams of P across the slice is given by $(\mathrm{d}P/\mathrm{d}x)\,.\,\mathrm{d}x$. If P molecules take time, dt, to cross the slice the loss of material is $v\,.\,(\mathrm{d}P/\mathrm{d}x)_t\,.\,\mathrm{d}t$. Hence the total loss of material as monomer and polymer is given by the equation:

$$\mathrm{d}M+\mathrm{d}P = v\,.\left(\frac{\mathrm{d}P}{\mathrm{d}x}\right)_t \mathrm{d}t \tag{8.6}$$

Equation (8.6) was derived from an earlier discussion concerned with chromatography (de Vault, 1943).

Equations (8.5) and (8.6) can be solved to predict the forms of the Schlieren pattern (neglecting diffusion) that results from the polymerizing system. Differentiation of Equation (8.5) and substitution in Equation (8.6) leads to Equation (8.7) (bearing in mind that $x = 0$ when $t = 0$):

$$\left(1+\frac{nM^{n-1}}{K}\right)\frac{x}{vt} = \frac{nM^{n-1}}{K} \tag{8.7}$$

The symbol δ is used to replace x/vt so that Equation (8.7) is transformed to:

$$M = \left(\frac{K}{n}\cdot\frac{\delta}{1-\delta}\right)^{\frac{1}{n-1}} \tag{8.8}$$

If Equation (8.5) is now used it is possible to write an equation for P by substitution in Equation (8.8):

$$P = \frac{K^{\frac{1}{n-1}}}{n^{\frac{n}{n-1}}}\left(\frac{\delta}{1-\delta}\right)^{\frac{n}{n-1}} \tag{8.9}$$

Summing Equations (8.8) and (8.9) leads to Equation (8.10) for $M+P$ which when differentiated gives Equation (8.11) for the form of the Schlieren peak:

$$P+M = \frac{K}{n}\left(\frac{\delta}{1-\delta}\right)^{\frac{1}{n-1}}\left\{1+\frac{1}{n}\frac{\delta}{1-\delta}\right\} \tag{8.10}$$

$$\frac{\mathrm{d}(P+M)}{\mathrm{d}x} = \frac{1}{vt}\cdot\frac{1}{n-1}\left(\frac{k}{n}\right)^{\frac{1}{n-1}\cdot\frac{2-n}{n\delta-1}}\left(\frac{1}{1-\delta}\right)^{\frac{2n-1}{n-1}} \tag{8.11}$$

It was assumed that the sedimentation coefficient did not vary in the above derivation. A study of Equation (8.11) reveals that a dimerization would lead to a single peak with a trailing edge that is diffuse, while for values of n greater than two there are two peaks, one sharp and one diffuse. Thus one perceives the chromatographic analogy for the case of non-linear absorption isotherms. Figure 8.3 shows the type of Schlieren diagrams, neglecting diffusion, for data on dimerization and hexamer formation in the case of α-chymotrypsin. The minimum in the hexamerization pattern can be shown by the methods of differential calculus to be situated at $\delta = (n-2)/3(n-1) = 0{\cdot}267$ which is in accordance with $n = 6$. Massey and coworkers (1955) calculated that the areas of P and M are equal when $P+M = 3{\cdot}5$ mg/ml. Then, if $\delta = 0{\cdot}267$ where the area of the first peak terminates, we find the area of the monomer peak is 1·75 mg/ml so that K can be calculated from Equation (8.10). Actually there is 2·3 times as much monomer by weight at this point.

The slow peak has a constant area for all protein concentrations above 1·75 mg/ml and only for these concentrations is the fast peak seen at all. Figure 8.4 shows Gilbert's calculated apparent percentage of polymer and the sedimentation coefficients obtained using the vertical bisector of area as a measure of the speed of the peak. Obviously, such abnormal behaviour of the areas with changes in concentration, their asymmetric shapes and the unusual variation of s with concentration are of diagnostic importance for the detection of interactions.

The dimerization produces a single peak which is asymmetrical, having its tail towards the meniscus (although diffusion will partially mask this).

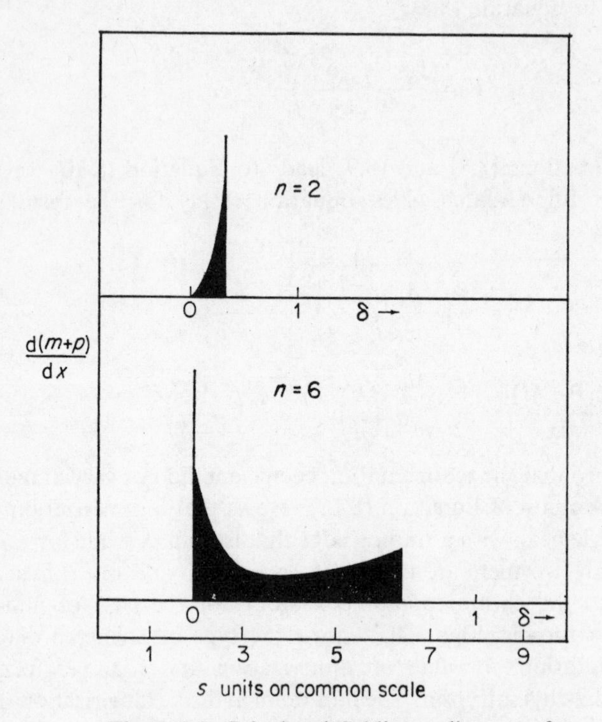

Figure 8.3. Calculated Schlieren diagrams for α-chymotrypsin in dimeric and hexameric cases when rapidly equilibrating with the monomer. From Gilbert (1955)

For $n = 2$, in the case of α-chymotrypsin, the observed sedimentation coefficients are markedly dependent on concentration, producing an increase in s value with concentration. This is quite contrary to the customary behaviour for non-aggregating systems (Figure 8.5). Cann and Goad (1968) have shown that for a mediated dimerization in which approximately 30 small trace molecules link two molecules of high molecular weight DNA a two peak system is to be predicted. This, of course, differs from the Gilbert single skew peak where there is not the mediation of the small molecules. Cann and Goad explained the pressure dependence of the two-peak system resulting from running the ultracentrifuge at different speeds as a Le Chatelier effect on the position of equilibrium.

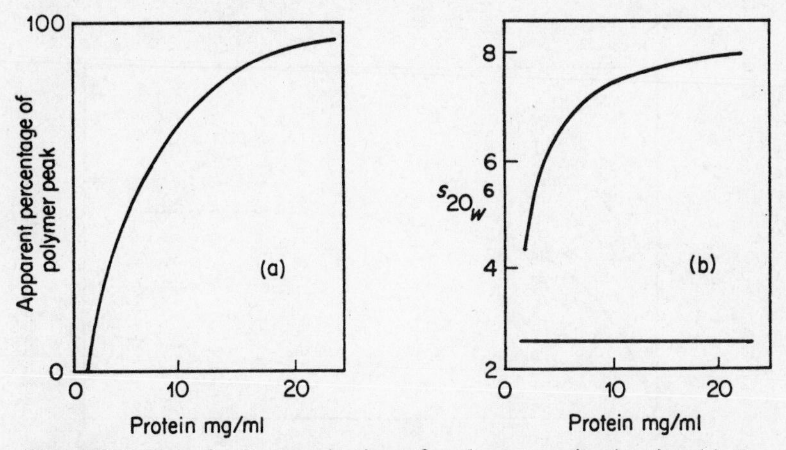

Figure 8.4. Data for hexamerization of α-chymotrypsin showing (a) the apparent percentage of polymer as area of diffuse peak, and (b) the variation of s values for the two peaks as determined from the centre of mass of the peak. From Gilbert (1955)

Even a weak tendency to dimerize may have a measurable effect on the observed s values. Gilbert (1960) has recommended the use of plots of relative velocity of sedimentation against relative concentration as an aid to diagnosis.

There have been many vindications of the Gilbert theory since its original conception. β-lactoglobulin was soon shown to conform under certain conditions of pH. For example, Gilbert and Gilbert (1962) have followed up work by Timasheff and Townend on this system where tetramers (β_4) can be formed or monomers ($\beta_{\frac{1}{2}}$). A recent study on glucose-6-phosphate dehydrogenase (Cohen and Rosemeyer, 1969) illustrates the modern approach. The enzyme is composed of sub-units of molecular weight 53,000 and of a single type, since hybridization is possible with the enzyme from erythrocytes of various mammalian species. The sub-units are aggregated to dimers, tetramers or octamers. The interaction between the tetramers is ionic in character, within the tetramer the dimers interact partly ionically and partly hydrophobically, while within the dimer the monomer–monomer interactions are hydrophobic. The changes from octamer to tetramer to dimer occur at near physiological pH. The dimer–monomer equilibrium is connected with binding of the coenzyme NADP. The variation of *s* values with ionic strength shows points of inflection that suggest changes in the structure of the enzyme. The weight average

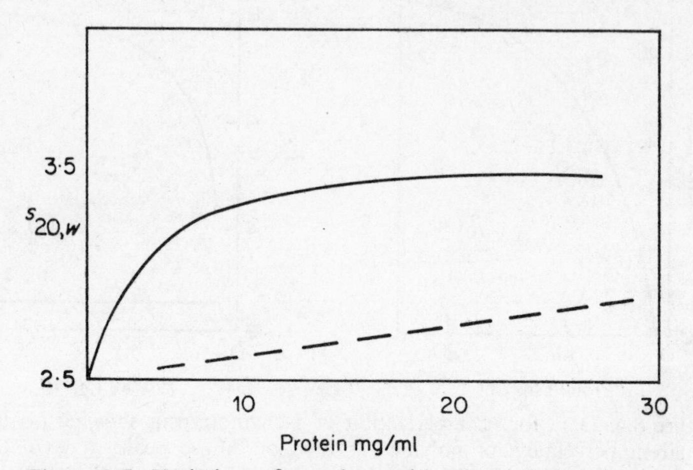

Figure 8.5. Variation of *s* values with concentration for the dimerization of α-chymotrypsin. The upper curve for chymotrypsin refers to $k = 1{\cdot}75$ mg/ml ($P+M = 3{\cdot}5$ mg/ml). The lower dashed curve relates to the expected behaviour for a dimerization with $k = 100$ mg/ml and $P+M = 200$. From Gilbert (1955)

molecular weights decreased when determined in buffer of different ionic strengths. Thus, a molecule of 210,000 molecular weight and $s_{20,w}$ of 9·0 Svedbergs changed to a half-molecule of 105,000 and 5·6 Svedbergs respectively. Skew patterns could be obtained under some conditions. Low salt concentrations at pH 6–7 could produce apparent molecular weights above 210,000, while reaction with maleic anhydride could produce the sub-unit of molecular weight 53,000. Molecular weights were determined at ionic strength 0·55 using the methods of van Holde and Baldwin (1958) to determine M_w and M_z which agreed at values of 208,000 and 207,000. Between ionic strengths of 0·5 and 2 M a single dissociation process occurred. If two molecular weights, M_1 and M_2, are involved the values of M_w and M_z are related according to the equation of Sophianopoulos and Van Holde (1964):

$$M_z = (M_1+M_2)-M_1M_2 \cdot \frac{1}{M_w} \tag{8.12}$$

Hence a plot of M_z versus $1/M_w$ should give a straight line from which the slope and intercept lead to a value of M_1 and M_2. Cohen and Rosemeyer were able to obtain values for the two molecular weights concerned of 105,000 and 210,000 respectively. Asymmetrical peaks could be produced in 0·1 ionic strengths with Tris-Cl buffers at pH 7·2 and pH 7·7 and in phosphate at pH 7·4, also in an ionic strength of 0·8 M when the pH was 6. These were interpreted on the boundary analysis of Gilbert (1959) and Fujita (1962).

The Gilbert theory given above is particularly valuable in emphasizing the physical ideas involved. It links the problem of polymerizing systems in the ultracentrifuge with known chromatographic theory. However, mention must be made of the work done by other workers on such systems.

Nichol and Bethune (1963) have given further thought to the application of the Gilbert theory to α-chymotrypsin. They point out that it is possible to determine $s_{20,w}$ (monomer) either by inhibiting polymerization by choice of medium or by extrapolating to zero concentration the weight average values obtained from the rate of movement corresponding to the square root of the second moment of the entire Schlieren diagram. However, it is difficult to determine $s_{20,w}$ (polymer) since its position is not defined. One can use Gilbert's expression, $\delta = (n-2)/3(n-1)$, but when n is not known an upper limit can be set approximately. When n approaches infinity the factor $(n-2)/3(n-1)$ approaches $\frac{1}{3}$ so that there is little sensitivity; for practical purposes, values of n equal to 10 or more can be considered infinite. For α-chymotrypsin a choice of $n = 6$ or 8 is arbitrary, but the hexamer was supported by other data. A value of $n = 8$ was supported by assuming that s^0 (polymer) $= s^0$ (monomer)$^{\frac{2}{3}}$ for a spherical shape, or even higher degrees of polymerization if shape factors are considered. In short, where n is known from other data the Gilbert theory enables one to find s^0 (polymer) which would otherwise be difficult to obtain.

Complexes

Gilbert and Jenkins (1956, 1959) studied systems of the type $A+B \underset{k-1}{\overset{k+1}{\rightleftharpoons}} C$ where the molecular weights are, respectively, M_A, M_B, and M_C and the concentrations of each in grams per litre are c_A, c_B, and c_C. It is possible to write rate equations:

$$\frac{d[A]}{dt} = -k_{+1}[A][B]+k_{-1}[C]$$

$$\frac{d[B]}{dt} = -k_{+1}[A][B]+k_{-1}[C]$$

$$\frac{d[C]}{dt} = +k_{+1}[A][B]-k_{-1}[C] \tag{8.13}$$

The sum of the three Equations (8.13) must equal zero to conserve mass. They can be written:

$$\frac{d[A]}{dt} = -k_{+1}\frac{c_A c_B}{M_A M_B}+k_{-1}\frac{c_C}{M_C}$$

$$\frac{d[B]}{dt} = -k_{+1}\frac{c_A c_B}{M_A M_B}+k_{-1}\frac{c_C}{M_C}$$

$$\frac{d[C]}{dt} = +k_{+1}\frac{c_A c_B}{M_A M_B}-k_{-1}\frac{c_C}{M_C} \tag{8.14}$$

Introducing new constants k'_{+1} and k'_{-1}, these equations take the form:

$$\frac{dc_A}{dt} = M_A(-k'_{+1}c_A c_B+k'_{-1}c_C)$$

$$\frac{dc_B}{dt} = M_B(-k'_{+1}c_A c_B+k'_{-1}c_C)$$

$$\frac{dc_C}{dt} = M_C(-k'_{+1}c_A c_B+k'_{-1}c_C) \tag{8.15}$$

Assuming a rectangular cell and neglecting diffusion the continuity equations become:

$$\frac{\partial c_A}{\partial t}+\bar{x}\omega^2 s_A \,.\, \frac{\partial c_A}{\partial x} = M_A(-k'_{+1}c_A c_B+k'_{-1}c_C)$$

$$\frac{\partial c_B}{\partial t}+\bar{x}\omega^2 s_\beta \,.\, \frac{\partial c_B}{\partial x} = M_B(-k'_{+1}c_A c_B+k'_{-1}c_C)$$

$$\frac{\partial c_C}{\partial t}+\bar{x}\omega^2 s_C \,.\, \frac{\partial c_C}{\partial x} = (M_A+M_B)(+k'_{+1}c_A c_B-k'_{-1}c_C) \tag{8.16}$$

Where $\bar{x} = (x_A+x_B)/2$. The theory is applicable where equilibrium is attained infinitely quickly (k'_{+1} and k'_{-1}) very large). Hence:

$$c_C = Kc_A c_B \tag{8.17}$$

where K is an equilibrium constant. The right hand sides of Equation (8.16) are then indeterminate but the equations can be handled when combined to give two equations:

$$\frac{\partial c_A}{\partial t}+\bar{x}\omega^2 s_A\cdot\frac{\partial c_A}{\partial x} = \frac{M_A}{M_B}\left[\frac{\partial c_B}{\partial t}+\bar{x}\omega^2 s_B\cdot\frac{\partial c_B}{\partial x}\right] \tag{8.18}$$

and

$$\frac{\partial c_A}{\partial t}+\bar{x}\omega^2 s_A\cdot\frac{\partial c_A}{\partial x} = -\left(\frac{M_A}{M_A+M_B}\right)\left[\frac{\partial c_C}{\partial t}+\bar{x}\omega^2 s_C\cdot\frac{\partial c_C}{\partial x}\right] \tag{8.19}$$

Equations (8.17), (8.18) and (8.19) are the basis of the solution for concentrations of c_A, c_B and c_C in terms of radial distance and time.

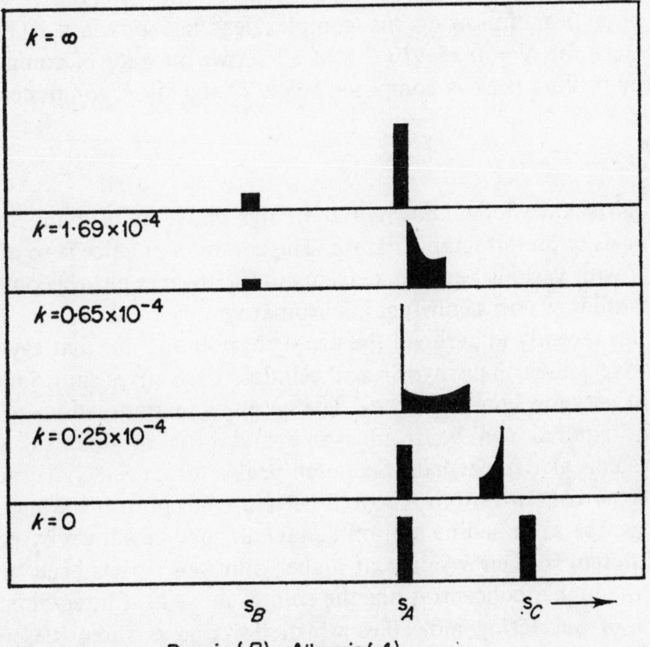

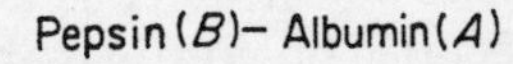

Figure 8.6. Sedimentation Schlieren patterns for a model simulating the pepsin–albumin system. From Gilbert and Jenkins (1963)

In simple cases, integration to find the boundary profiles may be performed analytically, but in more difficult cases numerical integration is employed. Gilbert and Gilbert (1965) have described the results to be expected in transport experiments as illustrated by an antigen–antibody reaction. Descending and ascending electrophoretic patterns were calculated and, by allowing a factor for the concentration dependence of the mobilities, a calculation of the Johnston–Ogston effect was possible. Thus we see that the theory of interacting systems can include the Johnston–Ogston effect and that its applicability extends beyond the centrifuge to electrophoresis and other techniques such as chromatography (cf. Nichol and Ogston, 1965). Gilbert and Jenkins (1963) have published their results for calculations on the pepsin–albumin system. Figure 8.6 shows the type of patterns to be expected for this $A+B \rightleftharpoons C$ system, neglecting diffusion but assuming a range of values of the equilibrium constant. A study of these Schlieren patterns is most illuminating. As the authors point out, the former assumption that the concentration of the slow component is approximately equal to the concentration of the component which is found as a peak following the complex leading boundary is not true. The pattern for $K = 0{\cdot}25 \times 10^{-4}$ moles/l shows no trace of component B while the trailing peak is composed solely of the faster component A.

The problem of gels

Biologists have long known that in the native state many materials exist in gels of an intractable nature. The common practice is to dilute the material with various aqueous reagents until enzymes etc., become amenable to study using centrifuges, chromatography and so on. Johnson (1968) has recently underlined the problem, pointing out that asymmetric macromolecules such as myosin and cellulose derivatives show interaction effects at very low concentrations. The necessary extrapolations to infinite dilution required can be troublesome and even of doubtful validity. Such systems give hypersharp Schlieren peaks and extremely curved plots of s against concentration. Above a certain concentration, which can be quite low, the same sedimentation behaviour may be shown by materials quite different in their s values at higher dilutions. It has been suggested that at the higher concentrations the solutes move as a three-dimensional network of interfering molecules which flows as a whole (described as 'plug' flow). Johnson (1968) reported observations on such a system in the hope that light would be shed on the kind of gel behaviour and sol–gel transitions that occur in the protoplasm. Agar and gelatin were used as model substances. The gel produces a somewhat sigmoid plot of log x against time. In the ultracentrifuge, the gel interface is sharp and clearly visible as an abrupt discontinuity. Gelatin shows a considerable amount of solute in true solution, sedimenting more slowly as a broad boundary than the gel interface; agar shows very little behind the gel boundary.

The gel interface is almost stationary at first, although it is subjected to the same centrifugal force that later causes an appreciable flow. Johnson suggested that the lag may be during a period when weaker bonds are being broken, ultimately to an extent that permits flow to occur. Later on in the run the concentration has increased to a point where bond formation equals bond breakage, giving a linear portion of the log x against time plot. Eventually, concentration rises to the point where bond formation exceeds bond breakage and the movement of the gel slows down. Incidentally, the concept of the increasing concentration of gel during a run is contrary to the customary radial dilution law that applies when solute is deposited on the cell bottom. Studies were made at various temperatures, ionic strengths and pH values. A general interpretation of gel behaviour in terms of general intermolecular attraction is possible, but much more work is required on such systems which are of great importance in the biological world.

REFERENCES

Cohen, P. and M. A. Rosemeyer (1969) *Eur. Jln. Biochem.*, **8,** 8.
Cann, J. R. and W. B. Goad (1968) *Adv. Enzymol.* **30,** 139.
De Vault, D. (1943) *J. Am. chem. Soc.*, **65,** 532.
Doty, P. and J. T. Edsall (1951) *Adv. Protein Chem.*, **4,** 35.
Fessler, J. H. and A. G. Ogston (1951) *Trans. Faraday Soc.*, **47,** 667.
Fujita, H. (1962) *Mathematical Theory of Sedimentation Analysis*, Academic Press, New York–London.
Gilbert, G. A. (1955) *Trans. Faraday Soc.*, **20,** 68.
Gilbert, G. A. (1959) *Proc. R. Soc. A*, 250, 377.
Gilbert, G. A. (1960) *Nature, Lond.*, **186,** 882.
Gilbert, G. A. (1963) *Proc. R. Soc. A*, **276,** 354.
Gilbert, G. A. and L. M. Gilbert (1962) *Nature, Lond.*, **194,** 1173.
Gilbert, G. A. and L. M. Gilbert (1965) *Biochem. J.*, **97,** 7c.
Gilbert, G. A. and R. C. L. Jenkins (1956) *Nature, Lond.*, **177,** 853.
Gilbert, G. A. and R. C. L. Jenkins (1959) *Proc. R. Soc. A*, **253,** 420.
Gilbert, G. A. and R. C. L. Jenkins (1963) in *Ultracentrifugal Analysis in Theory and Experiment* (Ed. Williams, J. W.) Academic Press, New York–London.
Green, N. M. (1969) British Biophysical Society, Physical Biochemistry Group Meeting, *Equilibria in Macromolecular Interactions*, London.
Johnson, P. (1968) The Chemical Society, London. Special Publication No. 23. *Solution Properties of Natural Polymers*, p. 243.
Johnston, J. P. and A. G. Ogston (1946) *Trans. Faraday Soc.*, **42,** 789.
Massey, V., W. F. Harrington and B. S. Hartley (1955) *Trans. Faraday Soc.*, **20,** 24.
McFarlane, A. S. (1935) *Biochem. J.*, **29,** 407, 660.
Nichol, L. W. and J. L. Bethune (1963) *Nature, Lond.*, **198,** 880.

Nichol, L. W. and A. G. Ogston (1965) *J. phys. Chem.*, **69,** 1754.
Ogston, A. G. (1953) *Trans. Faraday Soc.*, **49,** 1481.
Pedersen, K. O. (1936) *Nature, Lond.*, **138,** 363.
Pedersen, K. O. (1938) *C.R. trav. lab. Carlsberg, Ser. Chim.*, **22,** 427.
Schachman, H. K. (1959) *Ultracentrifugation in Biochemistry*, Academic Press, New York–London.
Trautman, R., V. N. Schumaker, W. F. Harrington and H. K. Schachman (1954) *J. chem. Phys.*, **22,** 555.
Sophianopoulos, A. J. and K. E. Van Holde (1964) *J. biol. Chem.*, **239,** 2516.
Van Holde, K. E. and R. L. Baldwin (1958) *J. phys. Chem.*, **62,** 734.

CHAPTER 9

Preparative ultracentrifugation

It is very likely that this chapter will be the first to be read; in fact, it may be the only chapter to be read by many. Most biologists will have used preparative centrifuges (later preparative ultracentrifuges) early in their careers and will have accumulated considerable practical experience. Noll (1969) writes bluntly of '. . . the prominent place and generally low standards of practice of ultracentrifugation among the techniques of protein biosynthesis . . .'. In the face of this provocative, but possibly true, statement the author diffidently suggests that the material given below will be supplemented by reading the review articles listed, including that by Noll (1969) himself. The author suggests that a knowledge of the analytical ultracentrifuge leads to a better understanding of the physical factors involved in using a preparative ultracentrifuge so this chapter, without hesitation, has been placed near the end. It has always seemed to be better to examine the raw material to be fractionated in an analytical ultracentrifuge first. In this way, using very little material, some of the main parameters such as approximate *s*-values and approximate composition are made apparent, and these enable the preparative runs to be planned on a more scientific basis. Now that analytical attachments are available for preparative ultracentrifuges (Beckman Instruments Inc., Martin Christ) at less than the price of a standard analytical machine it is possible that many more laboratories will be in a position to carry out analytical runs before preparative runs. There are many excellent reviews available on the subject of preparative ultracentrifugation; for example, Charlwood (1966) has surveyed the field of density-gradient ultracentrifugation, while Trautman and Cowan (1968) have covered preparative and analytical ultracentrifugation. The subject will be developed here as an extension of the analytical data to the preparative field.

Relationships between parameters

If it is assumed that a dilute solution and spherical particles are available, an equation can be developed relating molecular weight, solute density, solvent density, gravitational field and the velocity with which the particles will sediment. Equation (1.6) was obtained from the equation Force = frictional coefficient × velocity; restating this equation here:

$$\frac{M}{N}(1-\bar{v}\rho)x\omega^2 = f \cdot \frac{\mathrm{d}x}{\mathrm{d}t} \tag{9.1}$$

Assuming spherical particles, f, the frictional coefficient, becomes $6\pi\eta r$. The Stokes' radius, r, can be expressed in terms of molecular weight:

$$\text{*Volume of sphere} = \frac{4}{3}\pi r^3 = \frac{M\bar{v}}{N} \tag{9.2}$$

$$\text{Velocity of sedimentation} = \frac{\mathrm{d}x}{\mathrm{d}t} = \frac{M^{\frac{2}{3}}}{N} \cdot \frac{(1-\bar{v}\rho)}{\bar{v}^{\frac{1}{3}}} \cdot \frac{x\omega^2}{\eta} \cdot \frac{2\pi}{9} \tag{9.3}$$

Since $s = (\mathrm{d}x/\mathrm{d}t)/x\omega^2$, values of s could be used in this equation. Although assumptions have been made that are not likely to be fully met in a preparative run such as the high dilution many biological particles are compact in form if not actually spherical. Where there is appreciable asymmetry, the value of f will be increased, leading to a lower velocity than is given by Equation (9.3).

A study of Equation (9.3) brings out some very important practical points; the velocity of sedimentation increases as $M^{\frac{2}{3}}$ (for a sphere), and also with radial distance, x; it depends on the square of the speed of the rotor, ω^2, and is dependent on the buoyancy factor, $(1-\bar{v}\rho)$. The sign of the buoyancy factor can be reversed if $\bar{v}\rho$ is more than unity. The particle density $(1/\bar{v})$ is around 1·33 for a protein but can be as low as 1·02 for a lipoprotein. Hence, by increasing the density of the buffer, ρ, with added salt to around 1·06 it is possible to cause flotation of lipoproteins towards the meniscus. The viscosity coefficient, η, appears in the denominator, hence the velocity of sedimentation is decreased with an increase in this quantity. The coefficient of viscosity increases towards the bottom of the centrifuge tube when using density gradient techniques: in a special case, using an isokinetic gradient, the linear increase in x is offset by a linear increase in η and velocity of sedimentation is then independent of x.

* In Equation (9.2) there is an oversimplification in relating a Stokes' radius to an unhydrated partial specific volume. The equation derived (9.3) involves other assumptions as high dilution and is only suggested as a rough working rule.

Viscosity and density corrections

When relating s values obtained in an analytical cell, $(dx/dt)/x\omega^2$, to the movement of the same particles in a preparative run, the main parameter to correct for is the change caused in η by the new environment. The coefficient of viscosity is very dependent on temperature changes and on the presence of a material such as sucrose that may have been added for the preparative run. Density changes in the environment must also be considered but these are less dependent on temperature than the viscosity coefficient. Other parameters that may be different for the preparative run will be mentioned below, but a close examination of any change in the density and viscosity of the environment must receive first consideration. Fortunately, densities and viscosities are easily determined or can be looked up in tables. Tables of such data have been compiled and included in the important review of the gradient centrifugation of cell particles by de Duve, Berthet and Beaufay (1959). This review had a very important influence in making preparative centrifugation a precise branch of science rather than a mysterious art. Table 9.1 is an abridged version of the table compiled by de Duve and coworkers, which, when considered with Equation (9.3), should give an impression of the effect likely to result from a change in medium from water to sucrose solutions at various temperatures.

Table 9.1. Density and viscosity data for aqueous sucrose solutions (abridged from de Duve and coworkers, 1959)

Density		*Viscosity (poises)* $\times 10^2$			*Concentration*
0°	20°	0°	5°	20°	*per cent w . w*
1·000	0·998	1·79	1·52	1·005	zero
1·020	1·017	2·04	1·72	1·13	4·92
1·040	1·037	2·41	2·02	1·31	9·68
1·060	1·056	2·91	2·44	1·55	14·31
1·080	1·076	3·57	2·97	1·85	18·80
1·100	1·095	4·45	3·67	2·24	23·15
1·120	1·115	5·67	4·63	2·76	27·37
1·140	1·134	7·38	5·98	3·47	31·47

From the table it is immediately apparent that viscosity changes have a profound effect on the rate of sedimentation when changing to the very popular sucrose medium and that in such a medium the effect of temperature can be very great. The high viscosity of cold sucrose solutions renders it resistant to disturbance when tubes are being handled. On the other hand, the high viscosity is a nuisance where it is required merely to increase density to affect the buoyancy factor because it then increases the duration of an experiment. However, sucrose is cheap, convenient, well-tolerated by biological preparations and can be removed by dialysis. Osmotic effects may also be apparent with some particles in the presence of sucrose affecting their shape etc.

Various alternatives to sucrose are available to produce density gradients either to prevent convection or to increase density for buoyancy effects. Heavy water does not produce as great an effect as expected because it exchanges with the hydration layer on the particle thereby increasing its density. Colloidal thorium oxide, polyvinylpyrrolidone, caesium salts, rubidium salts, potassium tartrate (for viruses), bovine serum albumin (for erythrocytes) have all been used (Charlwood, 1966). Sulpholane, trimethylphosphate and urea can be used for ribosomal RNA when particular properties are required, such as transparency to ultraviolet absorption at 260 mμ, or when fractions must be added directly to a scintillation mixture, or when there must be no interference with orcinol or diphenylamine reactions (Parish, Hastings and Kirby, 1966). However, sucrose and caesium salts remain the most commonly used substances for increasing the density of the solvent.

*Effect of asymmetry**

Equation (9.3) was derived assuming the frictional coefficient was that for a spherical particle, $f_0 = 6\pi\eta r$, but a greater frictional coefficient becomes necessary for asymmetric particles. Table 9.2 (from Svedberg and Pedersen, 1940) indicates the relative increase required in the factor for ellipsoids of revolution. Thus, from Equation (9.3) the speed of sedimentation would be decreased by the factor in the table, providing the other parameters were unchanged. This represents a rough working guide but a more precise approach depends on the knowledge of more parameters that may not be easily available, for example, the osmotic properties of the particle, wall effects in some kinds of centrifuges, concentration effects and so on. The review by de Duve and coworkers, (1959) deals with these matters more fully. The material already given here should enable the data from an analytical run to be applied to a subsequent preparative run when correcting for the main effects due to changes in temperature, density and viscosity. For a range of spherical particles of the same den-

* See page 144 for highly asymmetric particles.

sity it has been shown that the s values are related to $M^{\frac{2}{3}}$. For long particles, however, s becomes almost independent of length (see chapter 10).

Table 9.2. Frictional ratios for prolate and oblate ellipsoids of revolution (abridged from Svedberg and Pedersen, 1940)

Axial ratio	*Frictional ratio f/f_0*	
	Prolate	*Oblate*
1·0	1·000	1·000
2·0	1·044	1·042
3·0	1·112	1·105
4·0	1·182	1·165
5·0	1·255	1·224
7·0	1·375	1·326
10·0	1·543	1·458
20·0	1·996	1·782

The determination of s from preparative runs

This problem is the converse of the foregoing sections in many ways. Noll (1969) relates corrected $s_{20,w}$ values to the rate, dx/dt at which particle of density ρ_p sediments through a medium of density ρ and viscosity, η, as follows:

$$\frac{dx}{dt} = s_{20,w}\frac{x\omega^2}{\eta}\frac{(\rho_p-\rho)\eta_{20,w}}{(\rho_p-\rho_{20,w})} \tag{9.4}$$

This equation normalizes data to the temperature and viscosity of water at 20°C. However, when a sucrose gradient is being employed the solvent density, ρ, and viscosity, η, are varying. If it can be assumed that osmotic effects are absent and that size and shape are not affected by the sucrose gradient, an integration of Equation (9.4) can lead to the $s_{20,w}$ value but the use of an isokinetic gradient simplifies the calculation. Noll (1969) gives a detailed description of the preparation of such isokinetic gradients. He also provides extensive tables for the preparation of such gradients for different types of preparative rotors.

An alternative procedure is to determine s values by interpolation between markers of known properties. Such an approach can, of course, support methods based on direct calculation. Martin and Ames (1961) used enzymes as markers; Stanworth and coworkers, (1961) used dye-labelled proteins and Charlwood (1963) used radioactively labelled proteins.

Determination of particle density

The following method was described by Trautman and Cowen (1968). An approximate estimate of particle density can be made by using a series of centrifuge tubes (angle or swing-out). Each tube is half filled with a solution greater in density than the previous tube. Low-density solvent is layered onto each solution, then, lastly, the sample is placed on top. The rotor is centrifuged at top speed for eight hours at the required temperature. After the run, the top and bottom layers are analysed for activity. The particle density approximates to the mean between the highest density allowing entry of the particle and the lowest density preventing its entry. When densities are required for the solvent, the conventional method of weighing in a density bottle and comparing with the weight of water is not possible with small drops of liquid. A micromethod is based on the use of a density gradient in a cylinder formed between *m*-xylene ($\rho_4^{20} = 0{\cdot}86$) mixed with bromobenzene ($\rho_4^{20} = 1{\cdot}50$) to give a density of 1·2 and then layered over bromobenzene. After a brief stir the column is used the next day. Drops of the solution come to rest and their position is compared with drops of known density made from salt solutions. For a more detailed discussion, Ifft, Voet and Vinograd (1961) have considered the problem of the determination of density distributions and density gradients in binary solutions at equilibrium in the ultracentrifuge.

The correlation of activity with a component of a mixture

A typical problem that presents itself is the isolation of an active particle from a mixture of biological origin. For example, a biochemist may have a crude preparation of an enzyme and wish to purify it further. In practice, many techniques would be brought to bear on the problem but only the use of centrifugal force is considered here. Examination in the analytical ultracentrifuge may reveal a number of Schlieren peaks and their sedimentation coefficients will have been determined. In order to discover which peak carries the enzyme activity an analysis is made of the supernatants as a function of time. Analysis of the pellet is of little value since all components are piling up at the bottom of the cell before the Schlieren peaks have reached the bottom meniscus. The disappearance of activity from the supernatant is complete when the peak associated with that activity reaches the cell bottom. Thus, by gently bringing the rotor to rest after various times, the activity in the supernatant can be measured and the activity–time plot reaches zero activity at the time the peak responsible reaches the bottom of the cell. A standard analytical cell can be used in this way but there are also partition cells available. Using a partition cell the run is terminated when a particular peak has just passed the position of the partition. After the run the material above the partition may be analysed separately from that below. There are two types of partition cell; (1) the cell with fixed partition that has a filter paper resting on a perforated partition in the hope (not valid for all cases) that it will prevent mixing but not interfere with sedimentation (2) the moving partition cell that has a partition forced to the bottom of the cell against rubber springs during the run but which rises after the run to isolate the material in the upper part of the cell (see p. 80).

Information on the buoyant density of the particle is sometimes useful. This may be found by centrifuging the material in a density gradient. The active particle will come to rest where its density matches that of the solvent. Details of these methods are given below. If the active particle were a lipoprotein of buoyant density 1·02 or thereabouts it would probably pay to concentrate it near the meniscus by a preparative run (not a density gradient) in a buffer of density brought to, say, 1·15 by adding a salt such as potassium bromide. Most proteins have densities near 1·33 and would move towards the centrifugal edge. More subtle distinctions in particle densities would require the use of a density gradient.

Classification of methods for centrifugation

The foregoing sections have shown that sedimentation is a process influenced by particle size, shape and density and that there are, in addition, properties of the solvent system such as density, viscosity, and temperature that can modify the sedimentation rate and even change its direction to produce a flotation. It is customary to define the different methods available as follows:

(a) *Rate sedimentation*

This does not require a density gradient in the solvent and is the commonest type of experiment. Solutes that are initially uniformly dispersed sediment to the cell bottom in the order of their s values and the movement of the interface records the progress of the experiment. In the analytical ultracentrifuge the progress of the Schlieren peaks follows such a process.

(b) *Density gradient separations*

(1) *Rate zonal method.* A zone of solute is placed at the top of a cell or tube in which solvent has a preformed gradient of density. Components of

the mixture sediment in zones roughly in the order of their s values. The density gradient serves to prevent convection.
(2) *Isopyknic or preformed gradient isodensity method.* A preformed solvent density gradient spans the range of particle densities of interest. Particles come to rest in the gradient when their buoyancy term $(1-\bar{v}\rho)$ is zero. That is, they move up or down until their particle density matches that of the solvent.
(3) *Equilibrium isodensity method.* Both solute and solvent are initially uniformly distributed. Upon centrifugation the salts (e.g. CsCl) in the solvent assume a sedimentation equilibrium distribution which produces a density gradient. The macromolecule solutes then band themselves at their isodensity positions.

The above methods will be developed in greater detail below.

The importance of rotor and cell design

From the above discussion it now appears that the position of the active component in a mixture has been established relative to the sedimentation behaviour of the other components of the mixture. A classification scheme has also been given for the main types of ultracentrifuge preparative experiment that can be used to effect a separation. The time has come, therefore, to consider the types of rotors available and how they can best be used.

The analytical cell

Analytical cells are not normally used for preparative work because they have a very small capacity, but they can be pressed into service for micromethods. It is convenient to discuss them first since they are designed to give as nearly ideal conditions for sedimentation as possible. The cell is sector shaped, and, if correctly aligned in the rotor, should not allow the solute to strike the side walls; the drive mechanism is made as vibrationless as possible; the cell bottom has the correct curvature so that sediment piles up evenly on it. However, the rotor cools adiabatically as the rotor accelerates and the cell contents warm up a little so that, after reaching speed, the temperature gradients have to even out quickly if there is to be no convection. Analytical cells are small so temperature gradients can disappear quickly, but a scaled-up sector for preparative work is not so ideal unless the sample is introduced or removed while the rotor is actually at speed; we shall see below that this is possible in the case of Anderson rotors. Ultracentrifuge rotors are flexibly suspended so that very precise balancing is not required. However, the position of static balance does not coincide with that of dynamic balance so that rotors can wobble at low speeds. A stabilizing device was described by deDuve (1959) to steady preparative rotors during the critical period and an attachment is available on some preparative centrifuges for this purpose. (Beckman Instruments Inc.). Other makers claim that they obviate the difficulty of low speed wobble in the general design of the instrument. When accelerating and decelerating swirling effects occur in all kinds of cells.

Swing-out and angle heads

Most preparative rotors in use are of these types. Figure 9.1 shows the behaviour of swing-out and angle rotors with cylindrical tubes compared

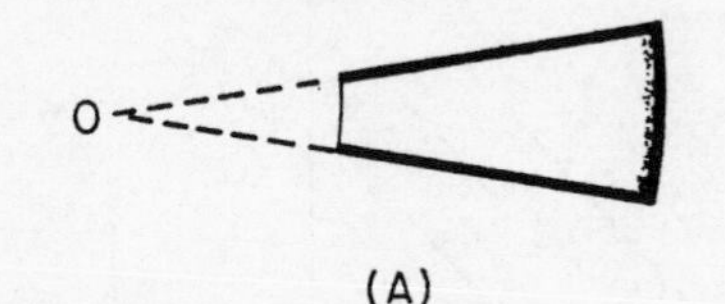

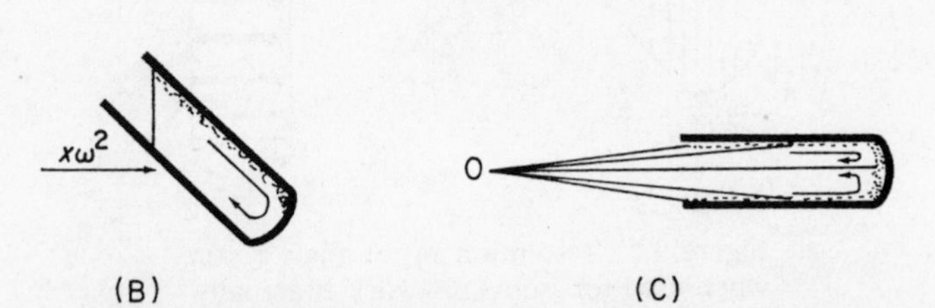

Figure 9.1. The effect of cell design on performance. The centre of rotation is at O and centrifugal force acts from left to right. (A) A sector-shaped cell is ideal for convectionless sedimentation. (B) An angle rotor produces strong convection by producing a dense layer against the outer wall. Deposition of pellet is very rapid since particles have a short distance to travel to the wall and convection assists deposition. This is not usually a good design for the separation of particles with similar sedimentation behaviour. (C) The swing-out type is usually preferred for difficult separations but particles strike walls to produce a heavy layer and subsequent convection. The arrows denote convection

with an ideal sector shape. Angle heads of type B give a very rapid collection of sediment because of the combination of a short path to the wall and the consequent convection caused by the heavy layer at the wall.

Such angle design was at one time considered only suitable for the separation of mixtures with widely spaced *s* values where speed and capacity were the prime considerations. For zonal work described below they are now considered to have very valuable properties.

Type C the swing-out rotor produces less convection than the angle rotor B and has been used very extensively for zonal experiments. Indeed, at one time all zonal runs were carried out in such rotors before the special virtues of the Anderson rotor and the angle head rotors for such runs were realized. Swing-out buckets produce undesirable swirling effects at the beginning and end of an experiment but the use of density gradients and controlled acceleration rates helps to offset this difficulty.

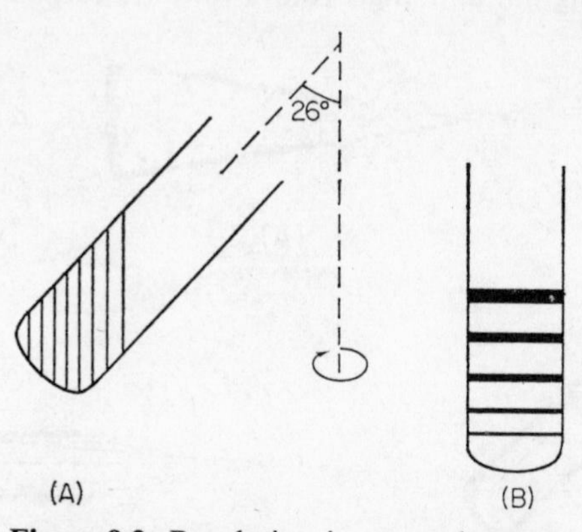

Figure 9.2. Resolution in an angle rotor when used for isopyknic work is actually improved when the tube is removed from its running position A to the sampling position B. From Flamm and coworkers (1969)

The Strohmaier cell (Strohmaier, 1966) fits into a swing-out rotor in place of the usual cylindrical tube but it has a sector-shaped cavity and small tubes are provided to enable samples to be removed at different levels for analysis. An optical device is also available for recording the solute zones as a function of height (Martin Christ).

More recently, it has been shown that the use of angle rotors for zonal work is possible and even has some advantages over the swing-out rotor. The increased danger of convection in the angle rotor is controlled by the sucrose density gradient. Flamm, Birnstiel and Walker (1969) have described the use of such a rotor for the preparation, fractionation and

isolation of single strands of DNA by isopyknic ultracentrifugation. After all, the gradients have to reorientate themselves through 90° when the rotor comes to rest and this is equivalent to the 90° reorientation of a swing-out tube (Fisher, Cline and Anderson, 1964). It is usually possible to obtain greater tube and rotor capacities in angle rotors. The gradient lengths are less and this speeds up the establishment of equilibrium. In the work of Flamm and coworkers, (1969) the RNA sedimented quickly and pelleted on the outer wall while there was better resolution of the DNA. Figure 9.2 from their paper illustrates the improved resolution claimed for the angle rotor.
The reorientation has the effect of constricting bands near the bottom of the tube but of expanding those near the top. In fixed angle rotors, band peaks are separated by greater volumes than in the comparable swing-out design so that mixing caused by disturbance is minimized. Larger samples can be taken without contamination. The authors were able to achieve comparable resolution in a swing-out rotor only by using one-tenth as much DNA per tube and by collecting four times as many fractions.

The Anderson rotor

An important advance in methodology was the development of the Anderson rotor (Anderson, 1965). This rotor is commonly known as the zonal rotor although, of course, any other rotor can be used for zonal work. Anderson points out that if the bottom of a centrifuge tube of the swing-out type is at a radius twice that of the meniscus, then approximately 25 per cent of the total particles will collide with the tube wall during complete sedimentation (Figure 9.1C). The particles which hit the wall may clump and sediment rapidly to the bottom, may stick to the wall, or may be reflected back into the solution. Indeed, de Duve and coworkers (1959) determined parameters relating to the behaviour at the walls for different biological particles. The Anderson rotor (Figure 9.3) employs sector-shaped openings in a cylindrical space and employs fluid lines to the centre and periphery so that the rotor can be filled and emptied while running. This overcomes problems of low speed rotor wobble and swirling effects associated with swinging buckets. Gradients can be introduced rapidly since they are stabilized by centrifugal force rather than the earth's $1g$ field. The zones can be displaced through an ultraviolet recorder, with a flow cell, by pumping in dense sucrose at the rotor edge after the run. Probably the most important single advantage of the rotor is the considerable scaling up of quantities that is made possible.

The first Anderson rotors were of types A and B. A type A rotor had large radius and short height while a Type B rotor was proportioned more like that shown in Figure 9.3. Type A was used at about 6,000 rev/min for the fractionation of large particles such as intact cells etc., while the Type B rotor ran at 40,000 rev/min and could fractionate material down

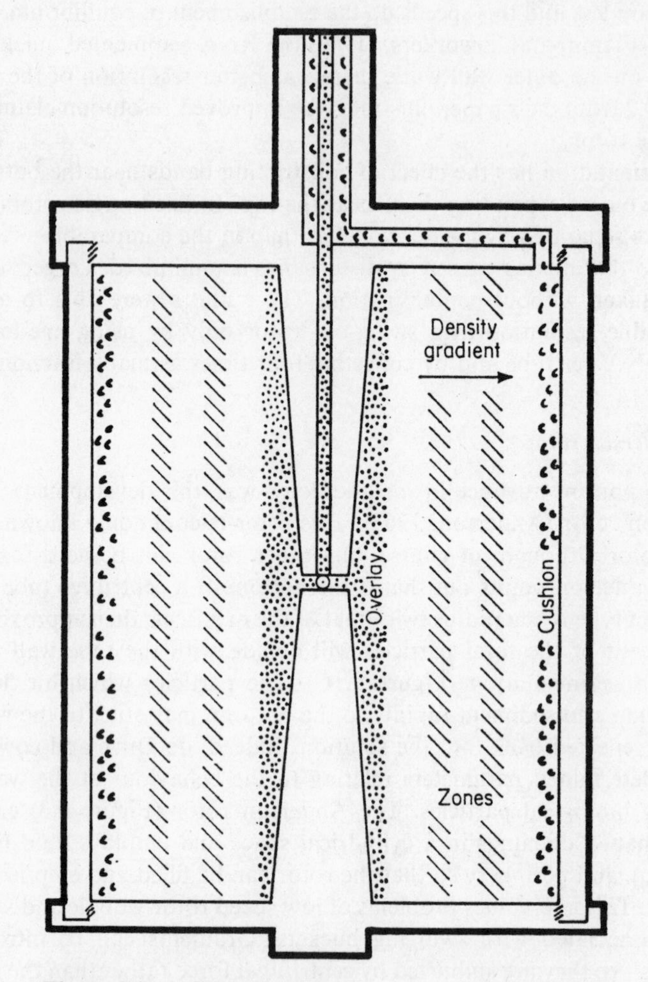

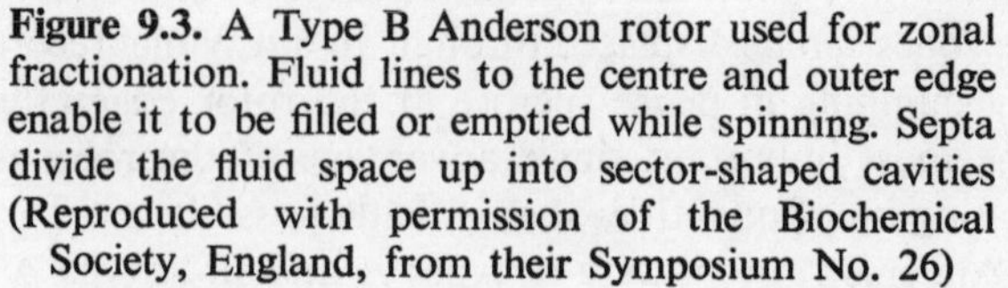

Figure 9.3. A Type B Anderson rotor used for zonal fractionation. Fluid lines to the centre and outer edge enable it to be filled or emptied while spinning. Septa divide the fluid space up into sector-shaped cavities (Reproduced with permission of the Biochemical Society, England, from their Symposium No. 26)

to 10 Svedbergs. Since then a large number of designs has emerged and the use of titanium has been exploited to obtain higher running speeds.

The development of Anderson (zonal) rotors for biologists was aided by the decision of the U.S. Energy Commission to foreclose the gas-centrifuge uranium-isotope activity of Electro-Nucleonics Inc. The Oak Ridge Molecular Anatomy (MAN) programme combined its know-how on zonal rotors with the firm in question and has produced a Model K rotor. This, at the time of writing, is probably the largest Anderson rotor with a capacity of 3·6–7 litres and able to process over 400 litres a day.

Reference to Figure 9.3 will clarify the method of operation of a typical Anderson rotor. The gradient is introduced from the rotor edge and is followed by a cushion of denser solvent to move the gradient towards the centre to exclude air. The rotor spins at a relatively low speed while being filled. When the sample is introduced at the centre of the rotor, a corresponding volume of the cushion solution is displaced. The sample is followed by an overlay of light solvent to move the sample zone a little way into the rotor, and the rotor is then accelerated to its operating speed. After the run, the rotor is decelerated to low speed and the rotor contents are displaced from the centre by pumping in dense cushion solution. The design of the rotor core sharpens each zone as it is pumped out. The Anderson rotor may also be used in a flow-through mode. Sample solution can flow through the rotor while particles of interest settle into a gradient from which they are fractionated and recovered. Difficulties with fluid line seals for high speed operation have led to the development of a rotor in which the fluid lines can be disconnected for the high speed part of the run and reconnected at low speed for the removal of samples.

Such rotors have an important industrial use. For example, the Type K rotor separates influenza virus up to 10 times purer than was possible with other centrifuges, with a corresponding decrease in the side effects caused by influenza vaccine. An isopyknic approach to the separation is used. Sucrose solutions provide the density gradient, although caesium chloride can be used in titanium rotors.

Selection of rotors and methods

Non-zonal rate method

If the s values are spaced well apart by a factor of approximately three or more a run in an angle rotor can achieve a useful separation. Such an experiment is particularly useful in the early stages of an isolation when large quantities have to be handled. Repetition of the procedure is often employed to achieve better resolution. Knowledge of the average s value for the particle, corrected for the preparative environment, leads to the problem of the selection of the rotor and the speed and duration of the run. Trautman (1963) has approached this as follows: The minimum s rate that a particle has to sediment from the meniscus to the bottom of some fraction removed from the tube is denoted by the symbol S^*_{min}. Consider any two levels, x_1 and x_2, then we can propose the following definition for S^*_{min}:

$$S^*_{min} = \frac{10^{13}}{\omega^2 t} \ln \frac{x_2}{x_1} \tag{9.5}$$

where S refers to Svedbergs and is first corrected for temperature and solvent from the analytical data. It is convenient to express time in hours and to make use of $\overline{ST}$ so that Equation (9.5) takes the form:

$$\overline{ST} = \frac{10^{13} \ln (x_b/x)}{(2\pi)^2 \,.\, (\text{rpm})^2_{max}} \tag{9.6}$$

where x is the radius to any point and $(\text{rpm})_{max}$ is the maximum rated speed for the rotor. The use of the function $\overline{ST}$ is possible, in spite of geometrical complications due to differently shaped tubes in angle or swinging bucket rotors, so that Equation (9.6) becomes:

$$\Delta\overline{ST} = (\overline{ST})_1 - (\overline{ST})_2 = \frac{10^{13} \ln (x_2/x_1)}{(2\pi)^2 \,.\, (\text{rpm})^2_{max}} \tag{9.7}$$

Negative signs are omitted for convenience in expressing $\Delta\overline{ST}$. It follows that the minimum apparent s rate of a particle which moves between any two levels is:

$$S^*_{min} = \frac{\Delta\overline{ST}}{t''}\left[\frac{(\text{rpm})_{max}}{(\text{rpm})}\right]^2 \tag{9.8}$$

where t'' is expressed in hours. Values of $\overline{ST}$ can be tabulated for any swinging bucket or angle rotor with their various tube combinations. Trautman (1963) and Trautman and Cowan (1968) give these values for Beckman rotors with examples of the use of the $\overline{ST}$ function. Taking an example from the second paper (Trautman and Cowan, 1968): approximately 100 ml of serum is available and it is desired to pellet the 19S component. A Beckman Spinco No. 40 rotor is selected taking 13·5 ml per tube. Since maximum speed is being used, Equation (9.7) takes the forms $\Delta\overline{ST} = 120 - 0$ and $t'' = 120/19 = 6{\cdot}3$ hours. The figure of 120 was obtained from the table for the No. 40 rotor. Closely related to the $\Delta\overline{ST}$ function is a similar quantity, k, defined as follows:

$$k = \frac{\ln x_2/x_1}{\omega^2} \cdot \frac{10^{13}}{3600} \tag{9.9}$$

where x_2 = maximum radius and x_1 = minimum radius and ω is the top angular velocity in radians per second ($\omega = 0{\cdot}10472 \times \text{rpm}$). Table 9.3 gives values of k for some current Beckman rotors.

Table 9.3. Some k values for Beckman rotors (by courtesy of Beckman Instruments Inc)

Type	rpm_{max}	*Max g*	*No. of tubes*	*Tube size* in.	*Tube capacity* ml	*Rotor capacity* ml	k
High force for small samples—many samples							
65	65,000	368,400	8	$\frac{5}{8}\times 3$	13·5	108	45
50Ti	50,000	226,400	12	$\frac{5}{8}\times 3$	13·5	162	77
High force and big sample volume							
60Ti	60,000	361,300	8	$1\times 3\frac{1}{2}$	38·5	308	63
50·1	50,000	275,000	8	$1\times 3\frac{1}{2}$	38·5	308	94
42	42,000	205,700	6	$1\frac{1}{2}\times 4$	94	564	156
High force swing-out							
SW65Ti	65,000	420,000	3	$\frac{1}{2}\times 2$	5	15	46
SW56Ti	56,000	408,000	6	$\frac{7}{16}\times 2\frac{3}{8}$	4·5	27	55
SW50·1	50,000	300,000	6	$\frac{1}{2}\times 2$	5	30	59
6 and 4 bucket swing-out—optimum resolution, wider peak separation in rate zonal work							
SW41Ti	41,000	286,000	6	$\frac{9}{16}\times 3\frac{1}{2}$	13	78	123
SW40Ti	40,000	284,000	6	$\frac{9}{16}\times 3\frac{3}{4}$	14	84	137
SW36	36,000	193,000	4	$\frac{5}{8}\times 3$	13·5	54	157

Note: Rotor designs are constantly being improved so the manufacturer must always be approached for the latest information

Thus, a Type 40 rotor pellets 20S ribosomes in $t = k/S = 120/20 = 6$ hours, but a Type 50 TI with $k = 77$ takes $77/20 = 3{\cdot}9$ hours. When a rotor is run below the maximum speed the k value is increased to $k \times (\text{rpm}^2_{max}/\text{rpm}^2)$. The choice of rotor will also depend on its capacity as well as its resolving power. A larger capacity may save a second run and the increased resolution of a longer tube may be an advantage.

An alternative treatment is to give a preparative rotor a performance index, P_i (Giebler, 1958) defined as follows:

$$P_i = \frac{(\text{rpm})^2}{\ln (x_b/x_m)} \tag{9.10}$$

where x_b is the radius to the cell bottom and x_m the radial distance to the meniscus. Then:

$$S^*_{\min} = \frac{10^{13}}{(4\pi^2 t'')P_i} \tag{9.11}$$

and

$$S^*_{\min} = \frac{10^{13}}{60^2 \,.\, 980t''} \cdot \frac{x_b \ln(x_b/x_m)}{\mathrm{RCF}_b} \tag{9.12}$$

The value of $S^*_{\min}$ can, of course, be obtained from the converse of the above. The time, t'', at which activity just leaves the supernatant can be found experimentally and used in the above equations to compute $S^*_{\min}$. Any sedimentation that has occurred during the acceleration and deceleration period must be allowed for by adding two-thirds of the run up and run down times to the actual running time at constant speed. A more precise correction is possible where the apparatus is equipped with an automatic integrator to read out the integral of $\omega^2\,dt$; the major ultracentrifuge manufacturers can supply this device.

Zonal methods

So far only the moving boundary method has been dealt with. It is capable of handling large quantities of material but is only capable of giving clean fractions when the s values are well separated and when the process is used repetitively. The situation is analogous to an attempt to separate chalk and sand by stirring them up with water in a cylinder and allowing to settle under gravity: Only the slowest component can be isolated in a pure state.

A more refined approach is to separate the components as discrete zones as follows:

(1) *The rate zonal method*, or moving zone method, depends on the layering of the mixture at the centripetal side of the cell or tube and allowing the components to follow each other in the order of their s values. The s values in question are those operational in the solvent used. Since convection has to be prevented by the use of a density gradient in the solvent, the s values in question depend on the density, viscosity, osmotic effect etc., encountered by the components of the mixture as they descend through the gradient.

(2) *Isodensity or isopyknic methods* (a) with preformed gradient, or (b) the equilibrium isodensity method. A density gradient is required in the solvent and can be formed by techniques described below, or, it may be

allowed to generate itself by sedimentation equilibrium from a solution initially uniformly mixed. In either case the solutes, which can be initially uniformly distributed, will band to form zones where the particle density matches that of the solvent. Sucrose solutions are often used as mentioned above or caesium and rubidium salts can be used to produce densities as high as 2·0. Such methods can be applied in preparative rotors or in analytical cells, but the equilibrium isodensity method is confined to analytical cells. The band-forming centrepiece mentioned in an earlier chapter enables an analytical cell to be used for small scale studies.

Density gradients are produced by techniques which are now well established. Layering of solutions using a pipette can produce a series of decreasing density steps which merge into a smooth distribution either on standing or after a brief stir with a wire. This waiting period is tedious but non-linear gradients are easily made. Probably the most popular device uses two cylindrical chambers, one with dense solvent and one with an equal volume of light solvent. The dense solvent is stirred. The bottom of the two cylinders is connected so that light solvent can flow into the dense (stirred) solvent while an outlet from the bottom of the cylinder with dense solvent allows a centrifuge tube to be filled (Figure 9.4). A linear gradient is obtained if the two cylinders are of equal dimensions. The apparatus is commonly machined out of a transparent plastic. The valve serves to isolate the two chambers during the filling operations. Dense solution flows out first. Many other gradient devices are available and often employ two syringes, one with dense and one with light solvent, mechanically driven to produce gradients of various curvatures. Large versions are needed for the volumes required for Anderson rotors and are commercially obtainable (Beckman Instruments Inc; Phoenix Instrument Co; Measuring and Scientific Equipment Ltd).

Britten and Roberts (1960) showed that when the sample is introduced on a density shelf for rate zonal work a density inversion can occur unless the sample itself is introduced in the form of a gradient (see Figure 9.5). This precaution is often omitted but it is as well to remember that there is always the possibility of convection occurring in this way.

After zones have been established, a way must be found to isolate them after the experiment. Tube slicers have been popular when tubes were mainly made of a thin cellulose nitrate. A pointed blade was driven through the tube just below the zone which could then be removed with a pipette. Slices down to 1 mm thick could be obtained and spillage was no problem in a properly designed slicer. Tube puncturing is commonly practised using a hypodermic needle driven through the side of the tube just above the pellet. As the drops pass through the needle they are collected for further analysis. Centrifuge tubes are now being made of much thicker materials that are less easily punctured by a needle so that with these a dense sucrose solution is allowed to flow gently to the tube bottom

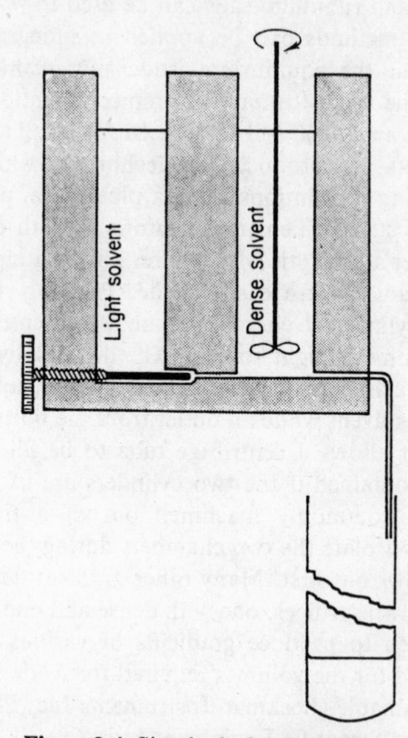

Figure 9.4. Simple device for making sucrose density gradients. The dimensions are chosen to correspond to the size tube being filled. The speed of outflow may be assisted by using a peristaltic pump. From Britten and Roberts (1960)

through a fine tube to displace samples out through another tube terminating just below a bung inserted in the top of the tube. Ultraviolet spectra of the zones are often required, and in these cases the outlow can be passed through a flow cell in a suitable ultraviolet spectrophotometer. Sucrose gradients have such high viscosities when cold that mechanical disturbance during the sampling operations does not give much trouble.

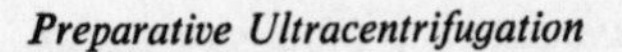

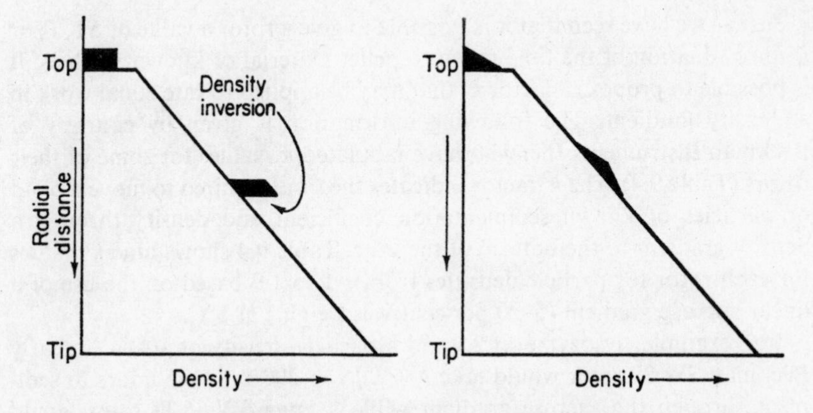

Figure 9.5. The drawing on the left shows how a density inversion can occur unless the sample is placed on the density shelf by a gradient device. From Britten and Roberts (1960)

Table 9.4. Values of k' (and k) for rate zonal work calculated for a sucrose gradient from 5–20 per cent at 5°C. (By courtesy of Beckman Instruments Inc).

Beckman rotor	*k′ factors for high performance rotors*									*k factors*
	Particle density									
	1·1	1·2	1·3	1·4	1·5	1·6	1·7	1·8	1·9	
SW 65 Ti	261	143	126	120	116	113	112	110	109	46
SW 56 Ti	313	170	150	142	137	134	132	131	129	55
SW 50·1	342	184	162	153	148	145	143	141	140	59
SW 41 Ti	689	379	335	317	307	300	295	292	289	123
SW 40 Ti	754	416	368	349	338	330	325	321	318	137
SW 36	894	491	435	411	398	390	383	379	375	157
SW 27	1492	816	722	683	661	646	636	629	622	265
Extra buckets for SW 27	1709	945	838	793	768	752	740	731	724	310
	k′ factors for earlier rotors (for comparison)									*k factors*
SW 50	390	211	187	177	171	167	165	163	161	68
SW 39	640	348	307	290	281	275	270	267	264	112
SW 25·3	2061	1145	1014	960	930	910	896	885	877	377
SW 25·2	1862	1024	907	858	830	813	800	790	783	335
SW 25·1	1881	1035	917	867	840	822	809	799	791	338

Just as we have seen that it is possible to give a rotor a value of $\overline{ST}$, P_i or k for evaluation of the time taken to pellet material of known s values, it is possible to propose a factor k' that may be applied to rate zonal work in a density gradient. The following information is given by courtesy of Beckman Instruments Inc. who have tabulated k' values for some of their rotors (Table 9.4). The k' factor indicates the time required to move a band of particles of known sedimentation coefficient and density through a density gradient to the bottom of the tube. Table 9.4 shows nine k' values for each rotor for particle densities from 1·1 to 1·9 based on the use of a linear sucrose gradient (5–20 per cent wet weight) at 5°C.

For example, lysozyme at 2S and a particle density of 1·4 in an early Beckman SW 39 rotor would take $t = k'/S = 290/2 = 145$ hours to sediment through the sucrose gradient while a later SW 65 Ti rotor would take $k'/S = 120/2 = 60$ hours. Where conditions are not similar with regard to gradient or temperature equations and tables have been provided by McEwen (1967).

The capacity of a density gradient

A problem of some practical concern is the capacity of a density gradient for carrying a zone. Equations suggested include those by Berman (1966), Svensson, Hagdahl and Lerner (1957), Gehatia and Katschalski (1959) and Vinograd and Bruner (1966). Spragg and Rankin (1967), using an Anderson rotor and sucrose density gradients resulting from a concentration range from 5–12 per cent sucrose, studied the capacity of the gradients for T_3 phage. They found that Berman's equation seemed to fit the experimental data. Berman was not concerned with the shape of the zone but merely considered how much material the density gradient could support. He made a simplifying assumption that for large particles the effect of diffusion on stability can be neglected, but the sectorial dilution of zones based on their radial position, the relative buoyant densities, the density gradient and viscosity of the supporting medium were considered. Berman's equation could be integrated by assuming the zone is narrow to give an equation equivalent to the one derived by Svensson and coworkers (1957):

$$m_{\max} = \frac{\pi L \rho_p (\mathrm{d}\rho_m/\mathrm{d}x)\bar{x}(\Delta x)^2}{(\rho_p - \rho_m)} \tag{9.13}$$

where

$m_{\max}$ is maximum amount of material held in the zone consistent with zone stability in the form of a Gaussian peak

L is the height of the rotor
ρ_p is the density of the particle
ρ_m is the density of the gradient at $\bar{x}$
$\bar{x}$ is the mean radial dimension of the zone
Δx is the zone width.

Good agreement was found between theory and experiment.

An important point is that there is a change in shape of the Gaussian peak when a zone crosses a discontinuity in the gradient and the shape of the peak is distorted when crossing this discontinuity. Furthermore, it is not certain how long the distortion produced when crossing the break persists when the peak is in the new smooth gradient. This is of practical importance when zones are started on a density shelf or in a region with a different density gradient from that experienced later. Asymmetry of a peak should not therefore be taken to indicate a hidden component unless a biological or chemical change can be proved across the peak.

McConkey (1967) has given a formula without derivation to estimate band capacity. His equation is as follows:

$$Y = \left(\frac{\mathrm{d}c}{\mathrm{d}h}\right) \cdot \frac{Ah^2}{2} \tag{9.14}$$

where Y is the total amount of the sample in mg that can be applied, h is the thickness of the layer of sample in mm, $\mathrm{d}c/\mathrm{d}h$ is the concentration gradient of the solute (RNA in this case on a 15–30 per cent sucrose gradient) in mg/ml/mm and A is the cross-sectional area in cm^2.

Overloaded gradients can exhibit droplet formation due to sucrose diffusing into a zone faster than the macromolecule can diffuse out. This effect can develop as the zones descend into regions of greater sucrose concentration. Intentional blurring of the lower edge of the zone can offset this difficulty.

REFERENCES

Anderson, N. G. (1965) *Fractions No.* 1, Beckman Instruments Inc., Palo Alto, Calif., U.S.A.
Berman, A. (1966) *Natn. Cancer Inst. Monogr.*, **21,** 41.
Britten, R. J. and R. B. Roberts (1960) *Science*, **131,** 32.
Charlwood, P. A. (1963) *Analyt. Biochem.*, **5,** 226.
Charlwood, P. A. (1966) *Br. med. Bull.*, **22**(2), 121.
de Duve, C., J. Berthet and H. Beaufay (1959) *Prog. Biophys. biophys. Chem.*, **9,** 325.

Fisher, W. D., G. B. Cline and N. G. Anderson (1964) *Analyt. Biochem.*, **9,** 477.
Flamm, W. G., M. L. Birnstiel and P. M. B. Walker (1969) in *Subcellular Components—Preparation and Fractionation* (Ed. Birnie, G. D. and S. M. Fox), Butterworths, London.
Gehatia, M. and E. Katschalski (1959) *J. chem. Phys.*, **30,** 1334.
Giebler, P. (1958) *Z. Naturf.*, **13b,** 238.
Ifft, J. B., D. H. Voet and J. Vinograd (1961) *J. phys. Chem.*, **65,** 1138.
Martin, R. G. and B. N. Ames (1961) *J. biol. Chem.*, **236,** 1372.
McConkey, E. H. (1967) *Meth. Enzymol.*, **12A,** 620.
McEwen, C. F. (1967) *Analyt. Biochem.*, **19,** 23.
Noll, H. (1969) in *Techniques in Protein Biosynthesis,* Vol. II (Ed. Campbell, P. N. and J. R. Sargent), Academic Press, London, p. 101.
Parish, J. H., J. R. B. Hastings and K. S. Kirby (1966) *Biochem. J.*, **99,** 19P.
Spragg, S. P. and C. T. Rankin (1967) *Biochim. Biophys. Acta*, **141,** 164.
Stanworth, D. R., K. James and J. R. Squire (1961) *Analyt. Biochem.*, **2,** 324.
Strohmaier, K. (1966) *Analyt. Biochem.*, **15,** 109.
Svedberg, T. and K. O. Pedersen (1940) *The Ultracentrifuge*, Clarendon Press, Oxford, England.
Svensson, H., L. Hagdahl and K. D. Lerner (1957) *Sci. Tools*, **4,** 1.
Trautman, R. (1963) in *Ultracentrifugal Analysis in Theory and Experiment* (Ed. Williams, J. W.), Academic Press, New York, p. 203.
Trautman, R. and K. M. Cowan (1968) in *Methods in Immunochemistry*, p. 81.
Vinograd, J. and R. Bruner (1966) *Biopolymers*, **4,** 157.

CHAPTER 10

Some applications of the analytical ultracentrifuge to biological problems

Each year many hundreds of papers are published describing work in which an analytical ultracentrifuge has been employed. Of these papers, the great majority will fall within the field of biology, and a large number of these will come within the province of the chemistry of proteins and nucleic acids. In this chapter, a short account is given of five of the more important topics in this latter field:

(1) Criteria of purity
(2) The sedimentation of highly asymmetrical particles
(3) The detection of conformational changes
(4) The determination of chain molecular weights
(5) The use of double-beam absorption optics, and the 'scanner'.

A long review could of course be devoted to even one of the topics. In appeasement of the furious specialist, whose excellent and highly skilled work has been ignored, let it be said that the aim is to introduce the reader to basic principles; to give a limited number of selected examples; and to suggest simple practical procedures which could reasonably be adopted by most scientists having access to an analytical ultracentrifuge.

Criteria of purity

The analytical ultracentrifuge is used extensively to characterize newly isolated proteins and other macromolecular species. It also serves the routine function of providing a check on the quality of preparations made by established techniques. In both cases, the research worker asks questions of the type:

(1) Does my preparation consist of a single macromolecular species, or is

it polydisperse?
(2) If there is polydispersity, is this, (i) static, caused by the presence of aggregates or impurities, or (ii) dynamic, caused by association–dissociation reactions?
This last question is considered elsewhere (Chapter 8). For the moment we will deal only with static systems, and with those methods by which we can best estimate the degree to which they are 'monodisperse'.

It is important to recognize that the term 'monodisperse' is used in an operationally defined sense. Circular though the argument sounds, an important truth is inherent in the statement 'monodispersity in sedimentation analysis means that all the molecules show the same sedimentation behaviour'. We must not expect to detect polydispersity with respect to some molecular property whose effect on sedimentation behaviour is essentially trivial, such as the degree of amidination of glutamate residues, or single amino acid replacements. For the detection of polydispersity of this sort, alternative techniques are available (gel electrophoresis, fingerprinting, see Colowick and Kaplan, 1967).

As previous chapters have shown, all experiments with the analytical instrument yield data in the form of a variety of parameters (Schlieren peak area, peak position, height of Schlieren trace, optical density, fringe displacement) which are functions of time and radial distance (velocity and approach-to-equilibrium experiments) or of radial distance only (equilibrium experiments). Analysis shows that for monodisperse systems, certain derived parameters are invariant with time or distance, thus providing criteria for the detection of polydispersity. In particular, for a single monodisperse sedimenting species, we can say that
(1) In a velocity experiment, the concentration change over the sedimenting boundary will decrease with time in strict accordance with the square-law dilution equation (Equation (5.8)), and that in a region ahead of the boundary, a plateau region will be present in which the concentration is invariant with distance at a given time.
(2) In an equilibrium experiment, the calculated molecular weight will be invariant with radial distance, if due allowance is made for any dependence of the molecular weight on concentration.

These two statements provide us with rigorous criteria of monodispersity, in the defined sense. Let us consider how these criteria would be applied in practice.

Analysis of the sedimenting boundary

Let us assume that a single, symmetrical boundary appears to be present. It is customary to 'correct for radial dilution' by the square-law formula,

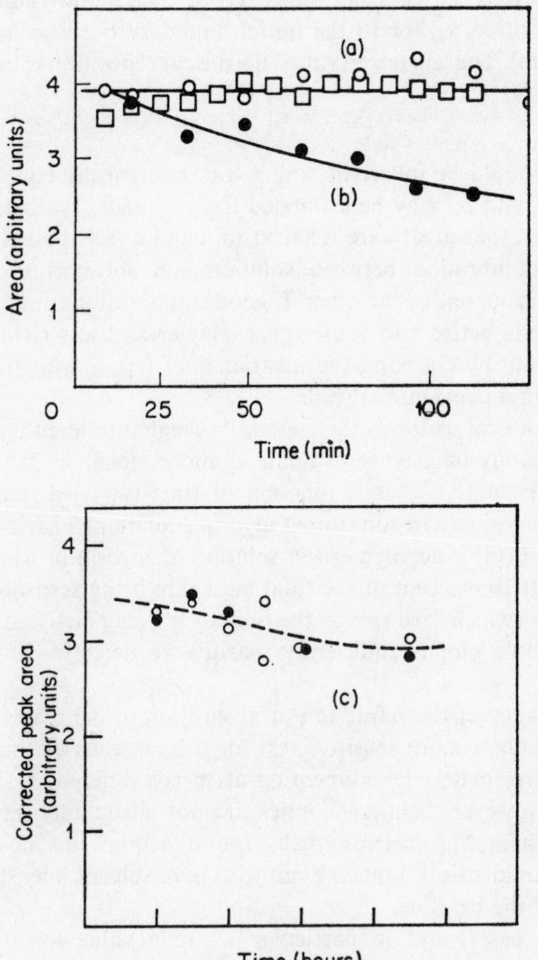

Figure 10.1. The corrected Schlieren peak area as a function of time for the muscle protein myosin (a) at low temperature (2·2°C) where the protein remains monodisperse (○); (b) at higher temperature (20°C) where the protein tends to aggregate (●); (c) in solution in 5 M guanidine hydrochloride, in which the protein is depolymerized to heavy chains (90 per cent total mass) and light chains. Data taken from Johnson and Rowe (1960) and unpublished work

and thus to refer all concentrations (c_t) at time t and radius x_t to the meniscus position x_m (or to the initial boundary position in a synthetic boundary run). The 'corrected value' for the concentration is thus given by:

$$(c_t)_{\text{corr}} = c_t\left(\frac{x_t}{x_m}\right)^2 \tag{10.1}$$

and $(c_t)_{\text{corr}}$ should be invariant with t, and numerically equal to c_m. The values for c_t and c_m may be estimated from a single, synthetic boundary run, provided that great care is taken to avoid a 'salt boundary', due to imperfect equilibration between solution and solvent, or to selective evaporation from one or the other. If good equilibration cannot be guaranteed, then it is better * to settle for a simpler and less stringent test of monodispersity by checking the invariance of $(c_t)_{\text{corr}}$ with time from an experiment in a conventional cell.

All three optical methods (Schlieren, Rayleigh interference and absorption) may readily be used to evaluate c_t and c_m (chapter 3). Figure 10.1 shows graphs of $(c_t)_{\text{corr}}$ as a function of time for (i) a 'monodisperse' solution of myosin, (ii) a solution of myosin containing heavier aggregated material; and (iii) a depolymerized solution of myosin in which the light chains (about 10 per cent of the total mass) are being separated from the heavy chains (which give rise to the boundary being analysed). It is clear that this simple plot is sufficiently sensitive to detect 5–10 per cent of impurity.

The constancy of the refraction or absorption in the plateau regions is almost certainly a more sensitive test for the presence of small amounts of polydisperse material, sedimenting at speeds different from the main boundary. However, Schlieren optics are not adequately sensitive here, and with both interferometric and absorption methods it is most important to perform an identical duplicate run with pure solvent only, to check the 'levelness' of the baseline.

In certain cases, and in particular where a value for the diffusion coefficient is known, an additional check on monodispersity is possible by analysing the spreading of the boundary, as suggested by Fujita (see Chapter 4). It is a pity that many workers (with small justification) regard the computation involved in this exercise as excessive.

Analysis of equilibrium experiments

'Our protein gave a linear plot of ln c against x^2, and is therefore shown to be homogeneous.' 'Our enzyme's molecular weight was found to be constant throughout the column, indicating homogeneity.'

* At least if Schlieren or Rayleigh interferences optics are in use.

In countless publications statements of this sort are made. Within their limits they are not wholly unreasonable statements; it is quite certain that a protein for which the ln c against x^2 plot shows a pronounced curvature is definitely *not* monodisperse. But for exact work this type of approach is both subjective and insensitive. It is subjective, because no defined criteria of 'straightness of line' are used; and insensitive, because the eye finds it difficult to detect slight curvature, even where a change in gradient (and hence M_{app}) of 10 per cent or more is involved.

A better approach is to evaluate the two molecular weight averages M_w and M_z, preferably from the same experiment. If Schlieren optics are used, this may conveniently be achieved by the use of Van Holde and Baldwin's method (Van Holde and Baldwin, 1958) to evaluate M_w at the 'mid-point, ($\bar{x}$) from:

$$k = \left(\frac{dc}{dx}\right)_{\bar{x}} \cdot \frac{1}{c^0} \cdot \frac{1}{\bar{x}} \tag{10.2}$$

and by the Lamm equation, using Method II of Rowe and Rowe (1970)* yielding M_z from:

$$\ln\left[\frac{1}{x} \cdot \frac{dc}{dx}\right] = \frac{k}{2} \cdot x^2 + \text{constant} \tag{10.3}$$

where k in both cases is defined by:

$$k = \frac{M(1-\bar{v}\rho)\omega^2}{RT} \tag{10.4}$$

If interference or absorption optics are used, then M_w is obtained from the normal plot of ln c against x^2, whilst M_z (and also M_n) may be derived using the methods of Teller and coworkers (1969) and Yphantis and Roark (1969).

In all cases, the ratio M_z/M_w will be equal to unity for a monodisperse system, whilst the degree to which it exceeds unity gives an indication of the degree of polydispersity in an inhomogeneous system.

The actual experiment should be a 'low speed' equilibrium run, not a meniscus depletion experiment. This is essential if all components of the system are to be conserved in that part of the cell in which measurements are made; otherwise grossly optimistic results can be given by a system containing heavy aggregates or impurities.

A major problem with both the above approaches—constancy of M_{app} throughout the cell and evaluation of the M_z/M_w ratio—arises when M_{app} is a significant function of concentration. (It will always be so to a certain

* It would be sounder in theory to determine M_z at the *mid-point*, by Method I of Rowe and Rowe (see Equation (11.1) but there would be a small loss in precision.

extent, see Equation (6.4)). The ratio M_z/M_w will presumably differ from unity even for a monodisperse system, although the mathematical analysis has not apparently been performed. In addition, the various graphical plots (e.g. ln c against x^2) will show curvature. Admittedly, this curvature will be opposite in sign to that produced by polydispersity, but how is one to hope to sort out a system where an apparently straight plot can result from opposite and compensating effects?

A recently developed method (Rowe and Rowe Method III) can be used here. These authors show that for a monodisperse system showing a linear dependence of M_{app} on concentration:

$$M_{app} = M_0(1-\alpha c) \tag{10.5}$$

where M_0 is the molecular weight at infinite dilution, then the following relationships hold:

$$\frac{{}^{(\frac{dc}{dx})_j, x_j}_{(\frac{dc}{dx})_i, x_i}\left[\ln\left(\frac{1}{x}\cdot\frac{dc}{dx}\right)\right]}{\int_{x_i}^{x_j}\frac{dc}{dx}\,.\,dx} = \frac{k_0}{2}\cdot\frac{{}^{x_j}_{x_i}[x^2]}{\int_{x_i}^{x_j}\frac{dc}{dx}\,.\,dx} - 2\alpha \tag{10.6}$$

$$\frac{{}^{c_j}_{c_i}[\ln c]}{{}^{c_j}_{c_i}[c]} = \frac{k_0}{2}\cdot\frac{{}^{x_j}_{x_i}[x^2]}{{}^{c_j}_{c_i}[c]} - \alpha \tag{10.7}$$

where

$$k_0 = \frac{M_0(1-\bar{v}\rho)\omega^2}{RT} \tag{10.8}$$

and

$$\alpha \simeq \sum\left(\frac{\partial \ln \gamma}{\partial c}\right) \tag{10.9}$$

for the solute components of the system.

Equation (10.6) is used with Schlieren optics, and Equation (10.7) with interference or absorption optics. The data, tabulated as dc/dx against x (Schlieren) or c against x (interference absorption) is differenced, taking in the general case the ith and j^{th} value of each parameter.

From the resulting straight-line plots, k_0 and hence M_0 can be evaluated from the gradient. This is useful, but more important for our present purpose is that the plots, being obtained by differencing, are found to be highly sensitive to the presence of polydispersity. Figure 10.2 shows two of the 'Method III' plots for a highly purified sample of ribonuclease, and for a slightly aggregated sample of lysozyme. The non-linearity in the latter case is very marked indeed.

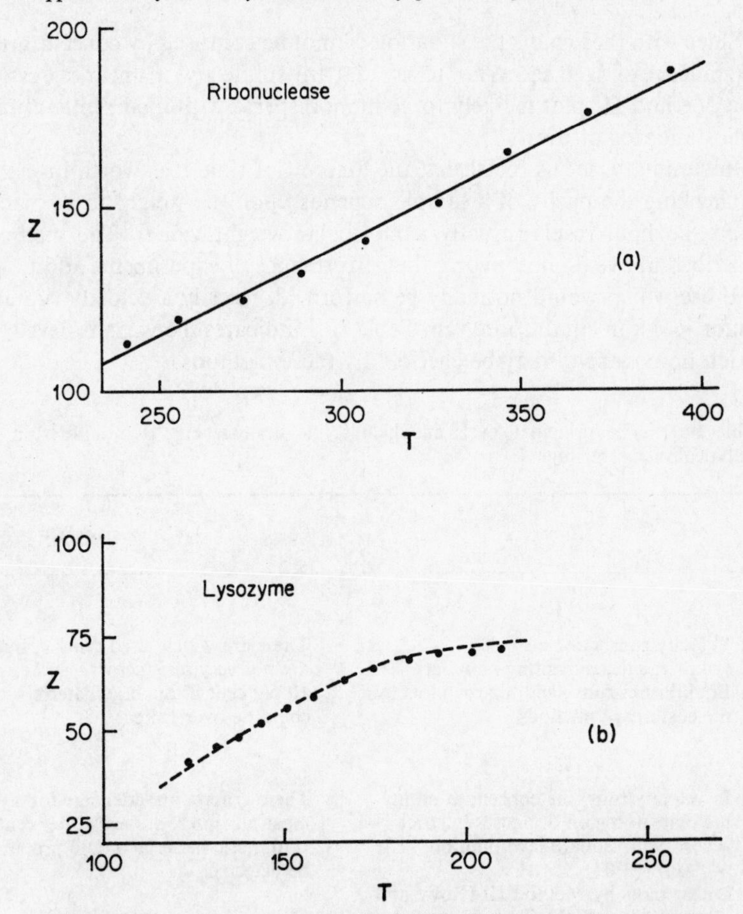

Figure 10.2. Two plots by 'Method III' of Rowe and Rowe. The chromatographically pure sample of ribonuclease gave a strictly linear plot by Method II (Equation (10.3)), over the entire cell. The Method III plot (a) is also linear, confirming the homogeneity of the sample. The coordinates are defined as:

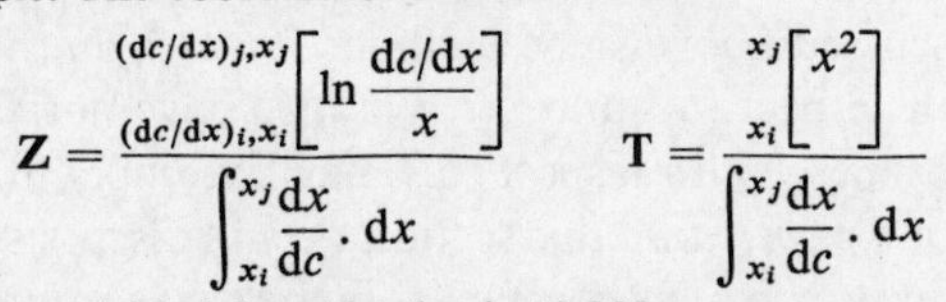

$$\mathbf{Z} = \frac{\left[\ln \dfrac{\mathrm{d}c/\mathrm{d}x}{x}\right]_{(\mathrm{d}c/\mathrm{d}x)_i, x_i}^{(\mathrm{d}c/\mathrm{d}x)_j, x_j}}{\int_{x_i}^{x_j} \dfrac{\mathrm{d}x}{\mathrm{d}c} \cdot \mathrm{d}x} \qquad \mathbf{T} = \frac{\left[x^2\right]_{x_i}^{x_j}}{\int_{x_i}^{x_j} \dfrac{\mathrm{d}x}{\mathrm{d}c} \cdot \mathrm{d}x}$$

The M_0 value yielded for this plot is 13,990.

The sample of lysozyme used showed a slight positive deviation from linearity, near the base of the cell. Data from this region were ignored in constructing the Method III plot (b); but even so the presence of polydispersity is detected by the marked curvature. M_0 values are found to be surprisingly insensitive to curvature in the Method III plot; in this case a reasonable value $M_0 = 14{,}200$ is yielded by a regression line taken from the points in (b)

Since with these plots the situation cannot be confused by concentration dependence of M, it seems fair to say that any solute giving linear regression in a Method III plot is likely to be monodisperse within currently attainable standards of purity.

In summary, let us recall that the amount of time it is worth investing in checking the purity of a sample depends upon the weight to be placed upon the final result (usually a molecular weight value). The methods described above do not involve the performance of experiments additional to those which would normally be performed, namely a velocity run and a short-column equilibrium run. Table 10.1 indicates the various 'levels' at which homogeneity may be checked by these methods.

Table 10.1. The principal levels at which solute homogeneity is monitored in the analytical ultracentrifuge

Stage	*Comment*
I	
(1) Velocity runs show a single, symmetrical sedimenting boundary (2) Equilibrium runs yield 'linear' plots by the customary methods	These are widely used criteria, but are not very satisfactory—some 10 per cent of dimeric material could be overlooked
II	
(1) In velocity runs, the concentration in the plateau region decreases in strict accord with 'square-law' dilution (2) $M_z/M_w = 1{\cdot}02$ (3) Linear plots by Method III (Rowe and Rowe, 1970) are obtained	These criteria are adequate for most purposes. Some 2–3 per cent of dimeric material would probably be detected

The sedimentation of highly asymmetrical particles

We have discussed above some methods for assaying the homogeneity of a solute with respect to s and M. What we have not considered is the fact that homogeneity with respect to s may not imply homogeneity with respect to M, or *vice versa*. In this section we will discuss systems where M may vary with approximately constant s; in the next section the significance of variation in s at constant M will be dealt with.

Consider the case of a rigid, elongated, polymeric molecule, of width $2b$ (fixed) and length $2a$ (variable). The molecular weight of any given molecule will be directly proportional to its length. We recall (p. 91) that for a homologous series of proteins, the sedimentation coefficient varies with $M^{\frac{2}{3}}$. Hence in this case s will vary with $a^{\frac{2}{3}}$. But with increase in length, the frictional ratio f/f_0 will also increase, in accordance with the equation (Perrin, 1936):

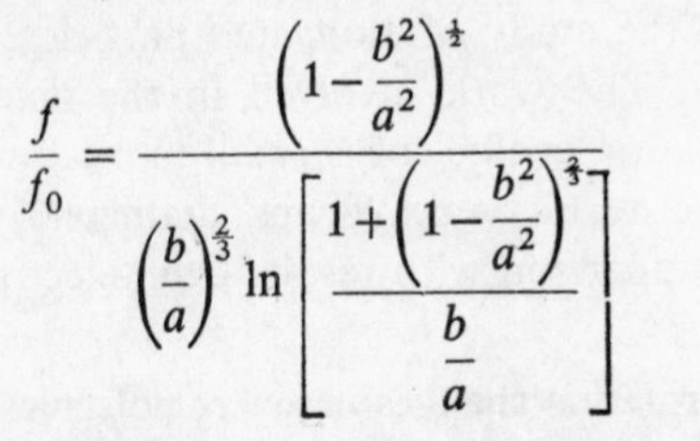

$$\frac{f}{f_0} = \frac{\left(1-\frac{b^2}{a^2}\right)^{\frac{1}{2}}}{\left(\frac{b}{a}\right)^{\frac{2}{3}} \ln \left[\frac{1+\left(1-\frac{b^2}{a^2}\right)^{\frac{2}{3}}}{\frac{b}{a}}\right]} \quad (10.10)$$

Inspection shows that for $b/a \ll 1$, at constant $b, f/f_0$ will vary inversely with $a^{\frac{2}{3}}$ (to an approximation; the change in the logarithmic term being relatively slow). It therefore follows that the sedimentation rate will change very little with length for highly elongated particles.

In passing, we may profitably note the need to test for weight polydispersity in such systems, using equilibrium methods, rather than relying on results from sedimentation velocity analyses.

The insensitivity of s to changes in length gives rise to a method for the estimation of the width of elongated molecules; combining Equations (1.6) and (10.10) (in approximated form), it can be shown that:

$$s_{20,w} = \frac{0{\cdot}222(1-\bar{v}\rho)}{\bar{v}\eta} \,.\, b^2 \ln\left(\frac{2a}{b}\right) \quad (10.11)$$

where ρ and η are the solvent density and viscosity of water at 20° (Peacocke and Schachman, 1954). Only a very approximate value for the axial ratio b/a needs to be assumed in order that b may be evaluated.

This method has been applied successfully to several systems, including nucleic acids (Peacocke and Schachman, 1954) and the protein F-actin (Johnson, Napper and Rowe, 1963). As an example of the insensitivity of the estimate to assumed axial ratio, it was shown that varying this over a four-fold range for F-actin (50 → 200) only caused the calculated width values to vary between 58·0 Å and 66·0 Å.

This value corresponds to the 'close-packed' or 'anhydrous' diameter of F-actin. It is one of the more useful features of this method that, although strictly speaking one should substitute the effective volume V_e (rather than $\bar{v}$) in the divisor in Equation (10.11) in order to yield the true molecular diameter in solution, simple algebra shows that the use of $\bar{v}$ in place of the (generally unknown) V_e leads to a value for the anhydrous diameter, at least in the absence of very strong specific interaction with individual solvent components. In other words, for highly asymmetrical particles, the sedimentation coefficient is directly proportional to the mass per unit length. As a consequence, it follows that changes in molecular diameter, at constant mass per unit length (i.e. swelling) will produce no change in the sedimentation coefficient.

As an approach to the study of elongated particles, this method has much to commend it. The work involved in the determination of an extrapolated sedimentation coefficient is small in relation to the information yielded; and the results (anhydrous diameters) lend themselves particularly well to comparison with results from electron microscopy.

The detection of conformation changes in macromolecules

We know from our basic equations (Chapter 7) that the sedimentation properties of macromolecules depend upon their shape and effective volume, as well as upon their mass. With the increasing interest in the dynamic properties of macromolecules, involving their ability to exist in a variety of conformational states of differing functional significance, it is interesting to consider how useful the analytical ultracentrifuge might be in this context.

To begin with, it is clear that the sedimentation velocity method can

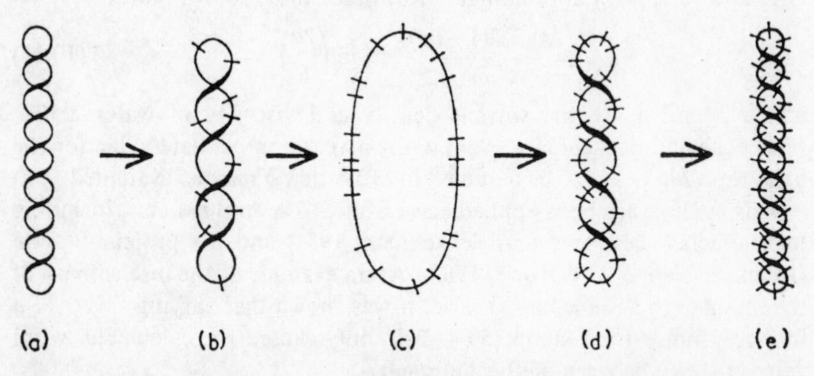

Figure 10.3. Diagrammatic representation (not to scale) of the removal, and reversal of supercoiling turns in polyoma DNA. The Watson–Crick helix is here represented as a single continuous line. The number of supercoiling turns in the original molecule (a) decreases as drug molecules (represented by bars perpendicular to the helix axis) bind to the DNA (b). The drug molecules are inserted at random. At equivalence, the accumulated untwisting due to the number of drug molecules bound just balances the initial number of supercoiling turns (c). The untwisting caused by the binding of further drug molecules leads to the introduction of tertiary twists now in the opposite, left-handed, sense ((d) and (e)). From Crawford and Waring (1967)

readily be used to detect fairly gross changes in conformation, even when an exact analytical treatment of the phenomena being studied cannot be given. A good example is the work of Crawford and Waring (1967) on supercoiled DNA.

It is known that polyoma virus DNA has an unusual structure, being both circular and supercoiled (see Figure 10.3a). It is thought that the super-coiling results from a deficiency of turns in the Watson–Crick double helix, of which each of the strands is composed. Certain drug molecules can 'intercalate' between the bases in the Watson–Crick helix, giving rise in successive stages to a compensation of this deficiency of turns, and then to an excess of turns, causing supercoiling in the opposite sense. Figure 10.3, a–e, shows these changes in diagrammatic form. In the absence of supercoiling, the DNA assumes an 'open-ring' form.

Now the hydrodynamic properties of 'open-ring' and of supercoiled helices are simply not understood, but a reasonable qualitative prediction can be made that the less compact, extended open-ring form will have a much greater frictional constant than the supercoiled conformation. The open-ring form should therefore sediment more slowly.

Crawford and Waring found this to be the case (Figure 10.4). The transition from supercoiled form to open-ring supercoiled form with successive additions of the drug molecules is very readily followed by the change in sedimentation velocity. In this case boundary sedimentation with absorption optics was used, and the results were analysed to yield estimates for the proportions of the various forms present. Figure 10.4 also shows that intercalation of drug molecules into the open-ring form of DNA—which will produce a simple, direct increase in circumference—causes a relatively slow decrease in sedimentation coefficient.

However, it is not only changes in large, extended molecules which are of interest to the biologist. Especially in the context of 'allosteric' phenomena, much attention is currently focused on the possibility of the larger enzyme molecules being able to exist in a series of differing conformational states, depending on the presence of various 'effectors'. In the context of metabolism, this means that the enzyme molecules may be 'switched-on' or 'switched-off', depending upon the needs of the organism. Can we use the ultracentrifuge to monitor a 'switching-off' process?

To answer this question, we must consider the types of conformational change most likely to occur, and then calculate the expected change in sedimentation behaviour. Only velocity studies will be considered, these being almost universally and exclusively used in practice. However, it should be observed that in principle equilibrium studies could also be used, not, of course, to monitor the molecular weight (assumed constant) but to follow any changes in the dependence of M_{app} upon concentration; this dependence is known to be a function of the excluded volume due to the particles, and hence of their asymmetry and molecular volume.

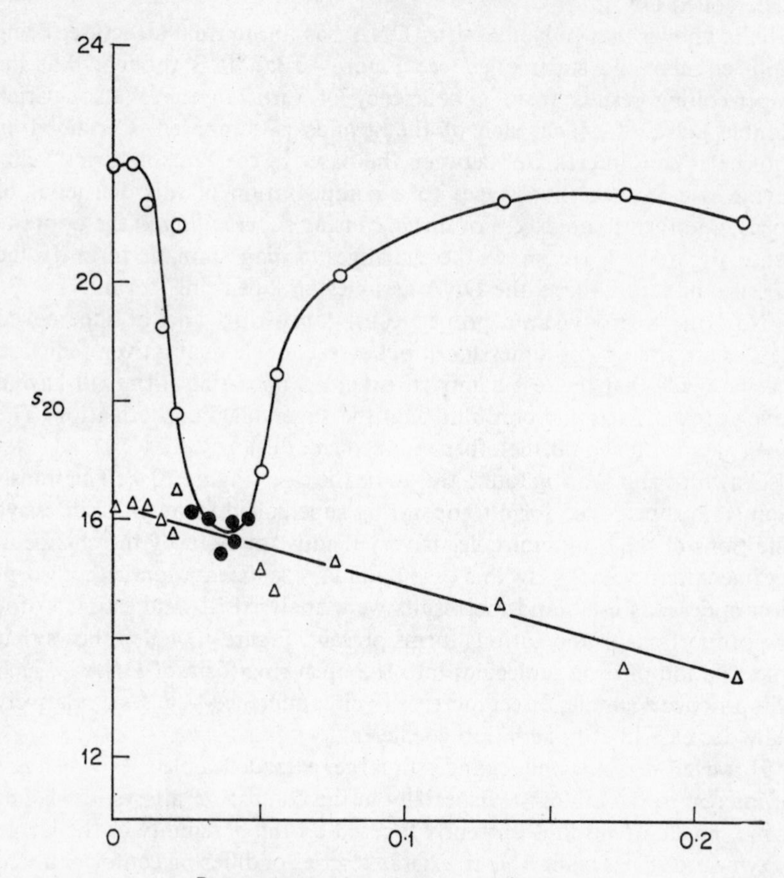

Figure 10.4. Effect of ethidium bromide on the sedimentation velocity of polyoma virus DNA. Samples of polyoma virus DNA, containing both fast (supercoiled) and slow (open ring) forms of the DNA, were exposed to a range of concentrations of ethidium bromide. Each sample contained 9 μg of DNA, plus ethidium bromide to give the binding ratio shown, in 0·5 ml Tris buffer (0·05 M, pH 8·0). The drug and the DNA were mixed and left at room temperature in the dark for 10 minutes before being centrifuged at 31,410 rev/min, 20°C.
—○—○—, Supercoiled molecules; —△—△—, open ring (unsupercoiled) molecules; —●—●—, only a single boundary was formed. From Crawford and Waring (1967)

When a molecule changes conformation, either its asymmetry (axial ratio), or its effective volume, or both, undergo concomitant changes. If the conformation changes are large, as in the case of protein denaturation reactions, then a marked change (usually a decrease) in sedimentation rate is observed. A variety of systems have been studied in this way, a recent example being pig liver esterase (Barker and Jencks, 1969) at neutral and acid pH (see Figure 10.5).

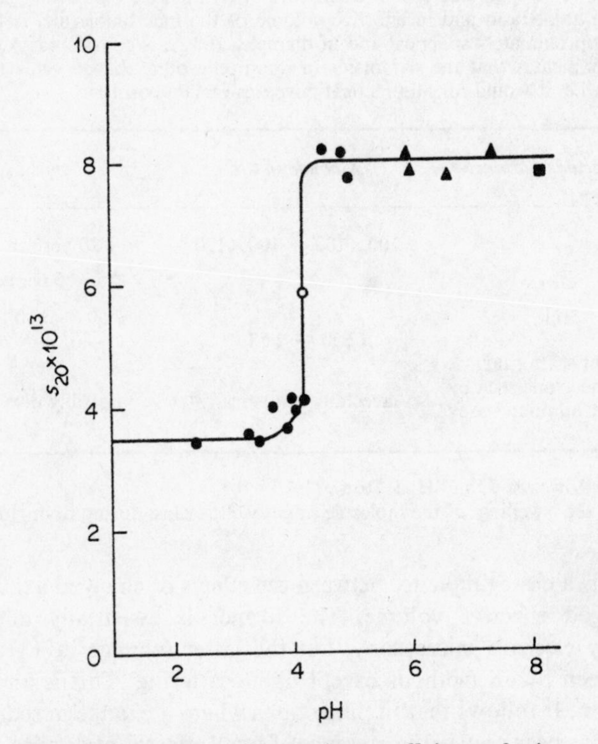

Figure 10.5. The sedimentation coefficient of pig liver esterase as a function of pH. From Barker and Jencks (1969)

It is unlikely that changes as big as this have any physiological significance, interesting though they are from a chemist's viewpoint. Most probably the 'allosteric enzymes', for example, only undergo changes of a few per cent in their axial ratio or effective volume. In Table 10.2 two cases are dealt with:

(i) A change of 20 per cent in axial ratio, viz. a/b increases from 1·0 to 1·2 (V_e constant)

(ii) A change of 20 per cent in V_e for a symmetrical particle, a/b remaining constant.

The resulting change in sedimentation coefficient is compared with the changes expected in parameters measured by other methods.

Table 10.2. A comparison of the sensitivities of certain macromolecular parameters to changes in axial ratio and in effective volume of the macromolecule. A hypothetical particle, approximately spherical and of diameter 100 Å, is considered. A dash (—) in the table indicates that the parameter in question would change value by less than 1 per cent, i.e. it would for all practical purposes remain constant

Molecular parameter	*Change in a/b*	*Change in V_e*
	$100\times100 \rightarrow 100\times120$	+20 per cent
$s_{20,w}$	—	$\Delta s = 6$ per cent
$[\eta]$	— (2·50 → 2·52)	2·50 → 3·00
Light scattering	—	—
Dimensions seen by electron microscopy	detectable statistically[a]	possibly detectable[b]

[a] Rowe, A. J. and H. J. Rowe (1970)
[b] the 'swelling' of the molecule might well reverse during drying down

There is a clear difference between the effects of changed axial ratio and of changed effective volume. The former is essentially undetectable, except by electron microscopy, but the latter (change in V_e) should be readily seen by all methods except light-scattering. This is an important conclusion. It follows that in those cases where a change in sedimentation velocity has been seen in the presence of small effector molecules, then those changes must have resulted from a change in V_e, regardless of whether a change in a/b also occurred or not.

Two examples may be quoted here. The enzyme aspartate transcarbamylase, isolated from *Escherichia coli*, is a large enzyme (M = 310,000) composed of four 'regulatory' sub-units and two 'catalytic' sub-units. In the presence of its substrate, carbamyl phosphate, which in this case case causes activation of the enzyme, Gerhart and Schachman (1968) were able to show a 3·6 per cent reduction in the sedimentation coefficient. This corresponds to an 11 per cent increase in effective volume, and suggests that the sub-units 'relax' in their packing to yield the catalytically active form of the enzyme.

Citrate synthase, from *Acinetobacter lwoffi*, is again a high molecular weight enzyme ($M = 240{,}000$), in this case of uncertain sub-unit composition. The coenzymes NADH and AMP act as allosteric effectors, being respectively the 'switch-off' and 'switch-on' signals. Rowe and Weitzman (1969) showed that 0·1 mM NADH caused a reduction of 7 per cent in the sedimentation coefficient, an effect abolished by the presence of AMP. In this case electron microscopy (by metal-shadowing techniques) showed an increase in axial ratio of the molecule; but negative contrast electron microscopy also showed a relaxation of sub-units to give a more expanded form, and in terms of our analysis above, we must suppose that it was this change which was responsible for the reduced sedimentation coefficient.

In attempting to measure possible changes in sedimentation coefficient, it is most important that proper controls be performed if spurious effects are not to be observed. Many of the substances used as effectors and substrates for enzymes exist as highly charged species in solution, and could affect sedimentation rates by charge effects alone.

One should therefore compare the system (enzyme+effector) with the system (enzyme+inactive effector analogue) and not with the system (enzyme alone). Gerhart and Schachman used potassium phosphate as an inactive analogue of carbamyl phosphate, whilst Rowe and Weitzman used NADPH for comparison with NADH. A full discussion of the possibility of artifactual effects is given by Gerhart and Schachman.

The determination of chain molecular weights

In studies on the large proteins, where each molecule is composed of a number of polypeptide chains, it is important to be able to estimate the molecular weights for the individual chains, as well as for the intact protein. The methods involved in separating and purifying the polypeptide chains do not lie within the scope of this book, but the end-product, the purified chains, must normally be dissolved in a strongly depolymerizing solvent to maintain them in a monodisperse form. The most efficient of such solvents is aqueous guanidine hydrochloride at strengths of up to 7 M.

A number of problems arise when attempts are made to determine molecular weights in this solvent. It is dense and viscous, leading to velocity and equilibrium runs of extended duration. It has a total refraction much exceeding that of the protein solute, and since redistribution in the cell is not negligible at the higher centrifuge speeds needed (chain molecular weights are obviously low in comparison with protein molecular weights), the meniscus levels require to be very carefully matched indeed if legitimate conclusions are to be drawn from experiments using double sector cells. Further, the partial specific volume of proteïns in this solvent will normally be unknown, and worst of all, it can be shown that the usual comfortable assumption of a 2-component system (water + buffer salt = 1 component) does not hold in this case. There can be very extensive selective binding of the solvent components by the protein.

This last complication is the most serious of all. Errors of 10 per cent or more in M values can be introduced if no connection is made. There are two alternative procedures here:

(i) Casassa and Eisenberg (1964) have shown that if in place of the term $(1 - \bar{v}\rho)$, the partial differential term $(\partial\rho/\partial c)$ is inserted in the equilibrium equations, then true 'anhydrous' molecular weights are yielded for the macromolecular component. The term $(\partial\rho/\partial c)$ must be determined experimentally, by comparison of the protein solution with solvent against which it has been exhaustively dialysed.

It is to be hoped that protein chemists will overcome their reluctance to attempt accurate density and concentration measurements, and will in future adopt Casassa and Eisenberg's method, especially as it also overcomes the problem of having to assume values for $\bar{v}$.

(ii) A correction for selective binding can be applied. An attempt can be made to correct for the preferential binding of a solvent component. The degree of binding can be estimated in a variety of ways. For example, Kielley and Harrington (1960) used equilibrium dialysis; Hade and Tanford (1967) used an isopiestic technique; Edelstein and Schachman (1966) performed a series of sedimentation equilibrium experiments, varying the solvent density by varying the concentration of guanidine hydrochloride, and extrapolating the quantity to zero, thus yielding an apparent $\bar{v}$.

The work involved in all these procedures is at least equal to that required for the estimation of $(\partial l/\partial c)$, but these and similar studies have been of great value in demonstrating that with all proteins it is the guanidine hydrochloride which is selectively bound, and not the water. The amount bound appears to vary from 0·01 to 0·17 g/g protein but it is possible that much of this variation is attributable to the poor precision with which the amount of bound water can be estimated. The sub-unit molecular weights, corrected for bound guanidine hydrochloride, are widely accepted as 'reasonable'; for example, the sum of the masses of the sub-units tallies closely with the mass of the individual, intact proteins. Significantly, Edelstein and Schachman's estimate for the mass of the aldolase sub-unit (50,000) is in contradiction with other available evidence, and theirs is the only study in which a selective binding of water has been claimed. It seems likely that the basis on which their extrapolation was performed (selective binding to be independent of concentration of guanidine hydrochloride) was not valid. For a full discussion, the reader is referred to Reisler and Eisenberg (1969).

Thus, where a new protein is under study, there seems little point in making a study of the amount of solvent selectively bound, unless the matter is of interest in itself. For highest precision, Casassa and Eisenberg's procedure should be followed. Otherwise, as a very good second best, Hade and Tanford's suggestion is reasonable; in this procedure, one must assume an apparent specific volume, ϕ', where

$$\phi' = \bar{v} - 0{\cdot}015 \pm 0{\cdot}005 \qquad (10.12)$$

where $\bar{v}$ is determined in ordering, aqueous solvents, and accept a ± 2 per cent uncertainty in M values.

The other obstacles to accurate working using guanidine hydrochloride as a solvent are less serious. The use of double-beam absorption optics (with the 'scanner') eliminates difficulties with the high refraction. The measurement of protein concentration in this solvent presents something of a problem, all the routine methods (Kjeldahl, Folin, biuret, ultraviolet absorption) being obviously inapplicable. A widely used and generally satisfactory procedure is to determine by control experiments the amount of solid guanidine hydrochloride required to be added per ml of aqueous solution, to give the final desired molarity, in measured volume. The extinction coefficient at 280 mμ can then be determined for one protein solution of known concentration, and this coefficient can be used thereafter. The solid guanidine hydrochloride must be stored in a controlled environment (a desiccator in a cold room is adequate). The molarities of guanidine hydrochloride solutions may be assayed from their densities (for tables see Kawahara, Kirshner and Tanford, 1965).

The use of double-beam absorption optics and the 'scanner'

This piece of instrumentation, described in Chapter 3, is probably the most important development in our field of the last decade. It offers the following advantages to the user:

(i) Very low concentration of solute can be employed (0·1 mg/ml for protein). In some cases the experimenter may not wish to take advantage of this facility since many proteins are less stable at high dilution, and handling of dilute solutions can result in difficulties from adsorption effects and heavy metal ion contamination. But in addition to the obvious application to enzymes available only in small quantity, the useful possibility now exists of studying an enzyme's sedimentation behaviour at about the same concentration as that at which its activity is assayed.

Further, at these low concentrations, the difference between M_{app} and M_0 (Equation (10.5)) becomes negligible for many globular proteins; in those cases where (Equation (10.5)) is large, the data can be handled using Equation (10.7) to yield M_0 under conditions which make Equation (10.7) a very good approximation (Rowe and Rowe, 1970). One of the more tedious tasks in protein chemistry, the measurement of seven or eight M_{app} values over a range of concentrations, is therefore rendered unnecessary.

Likewise, in sedimentation velocity analysis the use of very low concentrations permits investigations to be performed under conditions where interaction effects are minimal. Dissociation reactions may occur at these concentrations in certain systems; in the case of haemoglobin the depolymerization to give single-chain units (of weight comparable to myoglobin) was followed by Schachman and Edelstein, using a scanning system (Figure 10.6).

(ii) Types of experiment become possible which were previously outside the range of sedimentation analysis. By use of the monochromator, the system can be 'tuned in' to indicate the concentration distribution of a number of components, providing that these have distinctive spectral characteristics. The interactions of macromolecules with smaller molecules can therefore be studied. A good example here comes from the work of Steinberg and Schachman (1966). They showed that for the case of small molecules being bound to large molecules, the binding ratio (r, in moles/mole) can be evaluated from:

$$(A)_x = r \cdot (P)_x + A \tag{10.13}$$

where A and P are the concentrations of small and large molecules respectively, specified either for the original solution or at radial position, x. A plot of $[A]_x$ against $[P]_x$, each of which quantity is estimated at a definite wavelength, yields an estimate of r from the slope. Steinberg and Schachman performed a detailed study of the binding of methyl orange to serum albumin, and showed results comparable with those obtained by equilibrium dialysis. It seems likely that this type of approach will be increasingly used to study interaction of enzymes with substrates, inhibitors etc., and also, perhaps, to study the characteristics of enzymes in non-purified systems, if by the ingenious use of coloured, interacting small molecules these enzymes can be induced to leave a spectral 'foot-print' defining their presence.

(iii) The basic information output of the 'scanner' lends itself to analogue digital conversion, and hence to direct, on-line computation of results—perhaps even with feedback control of the instrument. Trautman (1966)

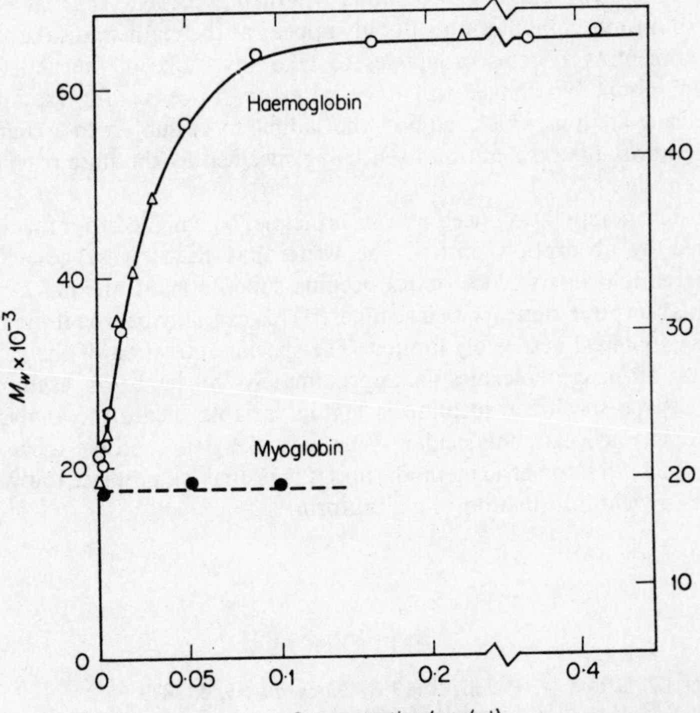

Figure 10.6. Molecular weights and sedimentation coefficients as a function of concentration for human oxyhaemoglobin and sperm whale myoglobin. The left ordinate gives the weight average molecular weight as determined by sedimentation equilibrium and the right ordinate represents the sedimentation coefficient in Svedbergs. The scale for the sedimentation coefficient was adjusted so that a value of 4·6 S corresponded to a molecular weight of 64,000 and the values of s were proportional to $M^{\frac{2}{3}}$. Sedimentation coefficients are indicated by Δ and molecular weights are represented by ○, ●. The concentrations are given on the abscissa in mg/ml. Data at concentrations below 0·05 mg/ml were obtained with light at 405 mμ and measurements with more concentrated solutions were made with light having a wavelength of 540 mμ. For the sedimentation equilibrium experiments speeds of 9,945 rpm to 47,040 rpm were employed. All sedimentation velocity experiments were conducted at 59,780 rpm. From Edelstein and Schachman (1967)

has given a review of the possibilities here. It should be borne in mind that the saving in time achieved by on-line computation of results will be very marginal if work is confined to currently accepted procedures. Most scientists, after all, already have access to computers and are happy enough to compile their data, if the computer will then process it. The real advantage of on-line computation will only appear as the mathematical analysis of sedimenting systems progresses to take advantage of the new-found 'elbow-room'. No longer will it be necessary to reject, for example, a promising analysis which requires the sample to be subject to a changing acceleration, the acceleration itself being specified by the state reached at a given time.

On-line computation need not, in principle, be confined to information yielded by absorption optics. The writer has heard it suggested that Schlieren and interference optics become superfluous if one has double-beam absorption optics with a scanner.* This can only be true if the range of one's interest is severely limited. The specific refraction of the various classes of macromolecules is approximately constant for each class, whereas the specific absorption is highly variable; hence the analysis of heterogeneous, macromolecular systems is likely to continue to be conducted by refractometric methods, and it is probable that these too will be made to yield information in digital form.

REFERENCES

Barker, D. L. and W. P. Jencks (1969) *Biochemistry*, **8,** 3879.
Casassa, E. F. and H. Eisenberg (1964) *Adv. Protein Chem.*, **19,** 287.
Colowick, S. P. and N. O. Kaplan (1967) *Meth. Enzymol.* (Ed. Hirs, C. H. W.), **11.**
Crawford, L. V. and M. J. Waring (1967) *J. mol. Biol.*, **25,** 23.
Gerhart, J. C. and H. K. Schachman (1965) *Biochemistry*, **4,** 1054.
Hade, E. P. K. and C. Tanford (1967) *J. Am chem. Soc.*, **89,** 5034.
Johnson, P., D. H. Napper and A. J. Rowe (1963) *Biochim. biophys. Acta*, **74,** 365.
Johnson, P. and A. J. Rowe (1960) *Biochem. J.*, **74,** 432.
Kawahara, K., A. G. Kirshner and C. Tanford (1965) *Biochemistry*, **4,** 1203.
Kielley, W. W. and W. F. Harrington (1960) *Biochim. biophys. Acta*, **41,** 401.
Peacocke, A. R. and H. K. Schachman (1954) *Biochim. biophys, Acta*, **15,** 198.
Perrin, F. (1936) *J. Phys. Rad.* (7), **7,** 1.

* It is regrettable that the Beckman Model E, when fitted with the 'scanner' and double-beam optics, can no longer be use with Schlieren or interference optics on a routine basis. The MSE Analytical has a clear advantage here, since all the optical systems can be selected in the complete instrument.

Reisler, E. and H. Eisenberg (1969) *Biochemistry*, **8,** 4572.
Rowe, A. J. and H. J. Rowe (1970) *Jl. R. microsc. Soc.*, **92,**
Rowe, A. J. and P. D. J. Weitzman (1969) *J. mol. Biol.*, **43,** 345.
Rowe, H. J. and A. J. Rowe (1970). *Biochem. biophys. Acta*, **222**, 647.
Steinberg, I. Z. and H. K. Schachman (1966) *Biochemistry*, **5,** 3728.
Teller, D. C., T. A. Horbett, E. G. Richards and H. K. Schachman (1969) *Ann. N.Y. Acad. Sci.*
Trautman, R. E. (1966) *Fractions, No.* 2, Beckman Instruments Inc., Palo Alto, Calif., U.S.A.
Van Holde, V. E. and R. L. Baldwin (1958) *J. phys. Chem.*, **62,** 734.
Yphantis, D. A. and D. E. Roark (1969) *Ann. N.Y. Acad. Sci.*

CHAPTER 11

Problems dealt with and problems to solve

In the last chapter, selected examples were given to show how the basic theory described in earlier chapters has been put to use in a variety of research projects. The reader will no doubt have projects of his own and may well feel that he faces problems different from those dealt with in the selected examples. Whilst the bibliography given should provide scope for further reading, it is true that most authors of scientific papers are concerned with the niceties of their own particular system, and the reader can have difficulty in extracting the pieces of generally relevant information.

The first part of this chapter therefore deals in brief fashion with a range of enquiries of the type which often come from centrifuge users. Many of the answers are based frankly upon the writer's own experience* and knowledge; no claims are made that every view expressed would win universal acceptance, and to maintain conciseness only limited justification is provided in most cases.

The second section ('Problems to Solve') is provided to give the reader a chance to try his own skill, or, alternatively, to suggest to the teacher material for class or examination use.

Problems dealt with

Sedimentation velocity experiments

Q. My protein is only stable at around 5°C. How can I correct the sedimentation coefficient to standard conditions (water at 20°C)?
A. You will have to accept some loss of precision in so doing. Firstly, determine the exact rotor temperature at which your experiment was performed (advisable at any temperature). The MSE system poses no problems, but with the Beckman Model E a procedure needs to be adopted

* With Phywe, Beckman and (currently) MSE Analytical Ultracentrifuges.

which is not widely known. The mercury pool contact, into which the thermistor needle dips, easily contaminates and gives spurious temperature readings. The following procedure eliminates any uncertainty on this score (Rowe, 1960).

1. At time t, read the apparent temperature of the rotor whilst it is in the chamber (T_1).
2. At time ($t+90$ sec) (work quickly!) read the true temperature of the rotor on the external stand, on which it has been calibrated (t_2).
3. At time ($t+180$ sec), having replaced the rotor in the chamber, re-read the apparent temperature (T_3).

The mean value $(T_1+T_3)/2$ gives an estimate for the apparent temperature at time ($t+90$ sec). If this mean value differs significantly from T_2 then the mercury pool is contaminated. If the velocity experiment has already been performed, then a correction equal to (T_2-T_1/T_3) should be applied to the estimated temperature.

You now have your experimental temperature T_e. Next, you will need to measure η and ρ for your solvent, at an exactly known temperature close (within 2°C) to T_e. The values of η and ρ for your solvent at T_e are then readily found, on the assumption that the viscosity and density of the solvent relative to water are constant over a small temperature range.

The correct values of $\bar{v}$ at 20°C and at T_e remain to be ascertained. If you are feeling heroic, go ahead, measure them both. For real accuracy, this is necessary. If you are prepared to settle for a little less precision, the value of $\bar{v}$ at 20°C (measured, calculated, taken from the literature or guessed at) may be 'corrected' for temperature dependence, assuming $\Delta\bar{v}$ for 1°C to be $6{\cdot}5\pm3\times10^{-4}$ ml/g/°C (based on data taken from Riethel and Sakura, 1963, and other sources). Note that the really important quantity is the temperature coefficient $\Delta\bar{v}$, and that this is not at all constant from one protein to another. The absolute value of $\bar{v}$ matters little, and indeed there is small point in measuring it. The reader may care to check that, assuming $\bar{v} = 0{\cdot}72\pm0{\cdot}03$ (for $\Delta\bar{v} = 6{\cdot}5\times10^{-4}$ ml/g/°C), introduces an error of only $\pm0{\cdot}4$ per cent in the correction factor, and hence in s, whilst the assumption for $\Delta\bar{v}$ renders s values uncertain to $\pm1{\cdot}5$ per cent. You can now apply Equation (4.4).

Q. Do I lose or gain anything in resolution by working at a lower temperature?

A. To an approximation, the resolution is independent of temperature, both sedimentation and diffusion rates being primarily sensitive to the viscosity changes in the solvent.

Q. How can I look for small, suspected changes in sedimentation coefficient?

A. Run both experimental and control samples in the same rotor, using wedge windows. Check the alignment of the wedge windows with great care—a few degrees of misalignment can falsify the radial distances for that cell. For a full discussion see Schumaker and Adams (1968).

Q. Can I start to estimate boundary positions as soon as the top of the Schlieren peak is clearly resolved from the meniscus?

A. No. Limited diffusion will render this estimate unreliable. Wait until the boundary is fully resolved from the meniscus. Likewise, avoid measuring up peaks which have almost hit the bottom.

Q. I wish to evaluate the sedimentation coefficient of an enzyme which is unstable in the presence of neutral salts. Do charge effects make this impossible?

A. It depends. The primary charge effect can be 'extrapolated out', working at low concentrations to avoid a long extrapolation. A cell with a long optical path-length would be advantageous here. The secondary charge effect does not extrapolate out, and trying to calculate its magnitude and allow for it is an uncertain business. So, if you can avoid having heavy ions like ATP around, you are probably in business; otherwise, think of another experiment.

Q. What is the best method for measuring up areas under curves?

A. Use a planimeter.* To produce an enlargement of the original film or plate a two-bath rapid processor is very useful as the paper, so long as it is not fixed, washed and glazed, is of adequate dimensional stability.

Approach-to-equilibrium experiment ('*Archibald method*')

Q. The multispeed procedure devised by Mueller (1963) seems to me to have a lot to commend it. Polydispersity can be detected, and M_w values, extrapolated both to zero time (original sample) and to infinite dilution, can be obtained. Why does no-one use it much?

A. Scientists like their beds as much as the next man does, and a method which requires the operator to be present to change speeds every 5 hours or so for two days is heading for serious consumer resistance problems. But as you say, scientifically speaking, the method has much to commend it. It would not be impossible for an ingenious person to automate the speed change, at least on the MSE instrument. The Model E would be more difficult to adapt, especially as the photographic system would also need modification.

Q. I am rather short of material, and doubt if I can perform a synthetic boundary run. Is there any way out of this?

* The Gelman planimeter is very suitable.

A. The obvious answer is to do a short-column equilibrium run instead of using the Archibald method. However, that experiment would take longer, and with an unstable sample might well be undesirable. If you have reason to trust the homogeneity of your sample, you can get a fair estimate for M by estimating $(\mathrm{d}c/\mathrm{d}x)_m$ at as early a stage as possible, and then accelerating the centrifuge to top speed. Correcting the area of the sedimenting peak for 'radial dilution' will then yield your required estimate for c^0.

Q. Where is the true meniscus position to be found?

A. The best position to take is the one which you find gives good results for 'standard' proteins—assuming the optics to be reasonably aligned.

Q. Is it worth estimating M values at the base of the cell?

A. No.

Short-column equilibrium experiments

Q. No matter which heavy oil I try, I always see an opaque layer (which I suspect is denatured protein) at the interface at the base of the solution column. Can this be avoided?

A. Very simply; by omitting the oil altogether. You will not now have a very good estimate for the effective column length, or be able to measure c or $\mathrm{d}c/\mathrm{d}x$ extrapolated to the base of the column. But with the 'mid-point' method having become obsolescent, the need for such estimates vanishes for most purposes.

Q. I wish to estimate the molecular weight of a polypeptide, approximate $M = 800$. Which is the best procedure?

A. Interference optics are out, because of the high speeds needed to study small molecules. You have a choice:

(a) The Archibald procedure (Schlieren optics)

(b) Short-column equilibrium, using Schlieren optics and evaluating by Equation (10.3) or (11.1) giving M_z

(c) Short-column equilibrium, using double-beam absorption optics and the 'scanner', yielding M_w from the normal $\ln c$ against x^2 plot. [If the peptide has adequate ultraviolet absorption.]

All three procedures should give a good estimate if carefully performed (± 2 per cent or better). However, as equilibrium will be reached very rapidly (1–2 hours, with a 1·5 mm column), the Archibald method offers no time-saving, and because of uncertainties in the meniscus position is best avoided.

Q. How about a small oligosaccharide?

A. With no worthwhile absorption regions, your choice would be (b) in the answer above.

Q. The protein which I am investigating aggregates very easily. Can I expect to obtain a reasonable estimate for the weight of the monomer?

A. The meniscus depletion method is, in principle, ideal in this situation. The use of Equation (10.7) will provide a check on the extent to which aggregates have been removed, as well as yielding M_0 if this has been done successfully. However, anomalies can arise, and all estimates should be treated with some caution (Jeffrey and Pont, 1969).

Q. I would like to determine the molecular weight of a small virus. Is this possible?

A. There are no theoretical difficulties, but there are instrumental problems to be overcome in working at the low speeds required (1,000 rpm for low-speed equilibrium; 2,000–3,000 rpm for meniscus depletion). For the Beckman Model E and MSE instruments, these are:

> *Model E.* The 'Electronic Speed Control' accessory is essential. The speed will then be accurately known and can be monitored by timing tachometer revolutions. The heavy rotor (An–J) must be employed to minimize vibrations from the drive unit. These vibrations, present even in newly installed drive units, and always present in ageing units, are the major obstacle to successful experiments.
>
> *MSE Analytical.* It is claimed by the makers (and borne out by the writer's limited experience) that the drive unit is essentially free of vibration at all speeds. It remains to be ascertained whether the speed control is adequate at low speeds. Periodic observation of the instantaneous read-out can be used to monitor the speed.

So, if you must use an ultracentrifuge (light-scattering gives good results in this size range) stick to meniscus depletion experiments and try to check your technique with a standard sample of known size (see Kado and Black, 1968).

Q. A sample of high molecular weight, polymeric material fails to give a trace at all in a low-speed equilibrium run. What has happened?

A. Has it gelled? Alternatively, an imperfectly sealed cell can give rise to evaporative convection (A. H. Cooper, personal communication).

Q. I have only a very small quantity of purified enzyme (< 1 mg). With no scanner available, which method will give me the best estimate for its molecular weight?

A. If you are skilled in the use of synthetic boundary cells, preparing a low speed equilibrium run by the method described by Charlwood (see Chapter 6, p. 73) will give the most information from a minimal sample.

Problems to solve

(1) The addition of mercuric ions to a solution of an enzyme increases its sedimentation coefficient from 3·0 *S* to 4·8 *S*. What is the most likely explanation for a change of this magnitude?

(2) A protein of well-defined sedimentation coefficient, $s_{20,w} = 6{\cdot}4$ *S*, $\bar{v} = 0{\cdot}73$ and intrinsic viscosity $[\eta] = 2{\cdot}3$ dl/g, is alleged to have a molecular weight of 400,000. Calculate a β factor for this protein, and decide whether or not the molecular weight value is a plausible one.

(3) A newly isolated polymeric protein appears in the electron microscope to consist of fibres some 200 Å in diameter. Their length is variable, but runs into thousands of Angstroms. Examination in the analytical ultracentrifuge shows a single peak, extrapolated $s_{20,w} = 35$ *S*. Are these findings compatible?

(4) The enzyme ribonucleotide reductase ($s_{20,w} = 9{\cdot}7$ *S*) polymerizes to an inactive form ($s_{20,w} = 15{\cdot}5$ *S*). How many monomers are likely to be present in the polymer? Is there any evidence of change in shape or effective volume?

(5) The data in the table below is taken from a meniscus-depletion experiment. Comment on the probable degree of homogeneity of the sample.

ln *n* (fringes)	x^2
1·892	46·77
1·965	47·25
2·068	48·08
2·153	48·81
2·264	49·63
2·379	50·40
2·513	51·28

(6) A very early publication (Svedberg and Sjogren, 1929) gave the following values for the hydrodynamic parameters of Bence–Jones protein: $s_{20,w} = 3{\cdot}7$ *S*; $\bar{v} = 0{\cdot}749$; *M* (equilibrium) = 35,000. On what basis might you be led to expect a degree of error in at least one of these values?

(7) A macromolecular complex, whose molecular weight is being investigated by the short-column equilibrium technique, is thought to undergo a slow, spontaneous dissolution to its components parts. Assuming this dissolution to have a positive temperature dependence ($Q_{10} = \sim 2$), would it be advantageous or disadvantageous to perform the equilibrium experiment near 0°C, rather than at 20°C?

(8) A solution of a protein (*A*), concentration 0·5 g/100, is mixed in a 1:1 ratio (approximately) with a solution of a second protein (*B*), also at

concentration 0·5 g/100 ml. The sedimentation coefficients of *A* and *B* (determined independently) are described by the regression equations:

$$A \quad s_{20,w} = 6{\cdot}80\ S - 3{\cdot}2\ c$$

$$B \quad s_{20,w} = 4{\cdot}15\ S - 2{\cdot}9\ c$$

where c is the protein concentration (g/100). If the mixture of *A* and *B* is centrifuged, will the relative areas of the peaks in the Schlieren diagram be an adequate representation of the exact relative proportions of *A* and *B* in the mixture?

(9) Plate VIII shows a Schlieren trace of a serum protein at equilibrium. Use the Method I equation of Rowe and Rowe (1970) to estimate its molecular weight from the data provided. The equation is:

$$M_x = \frac{\dfrac{d^2c}{dx^2}\,.\,x - \dfrac{dc}{dx}}{\dfrac{dc}{dx}\,.\,x^2}\,.\,\frac{RT}{(1-\bar{v}\rho)\omega^2} \tag{11.1}$$

at any position in the cell, distance x from the centre of rotation. What would be the most useful position at which to measure M, and why?

Answers to problems

Most of the problems could be discussed at some length. These answers give only the most important points which should feature in any real answer.

(1) Two possible answers present themselves: (i) the molecular weight of the enzyme is increased, either by polymerization or by binding of mercuric ions to the monomeric form, or (ii) the frictional ratio has decreased, at constant molecular weight.

The second answer is logically possible, but in practice is very unlikely. Few enzymes have a frictional ratio so high that it could be decreased in the proportion 3·0:4·8 without becoming less than unity.

Considering the first possibility, then, we recall that for a series of proteins of identical frictional ratio, the sedimentation coefficient is proportional to $M^{\frac{2}{3}}$ (Equation (7.18)). Dimerization should in general therefore increase s by $2^{\frac{2}{3}}$:1, i.e. 1·59:1. The observed change is in the ratio 1·60:1, which suggests very strongly that the correct interpretation of the data is that the enzyme dimerizes.

Again, the alternative explanation (binding of mercuric ions) cannot be dismissed solely on the basis of the data given. However, it would seem unlikely that the massive binding needed to increase the effective weight of each monomer by 59 per cent could occur without cross-linking of the monomers by the divalent mercuric ions.

(2) Use of Equation (7.18) leads to a value $\beta = 3{\cdot}5$. From Table 7.1 this corresponds to a prolate ellipsoid of axial ratio 200:1. Although large, this value cannot be called impossibly large. However, we can attempt to calculate the effective volume V of such a particle, from Equations (7.12) (to give v) and (7.15). The value yielded, $V_e = 0{\cdot}11$ ml/g, is very much less than the partial specific volume, a phenomenon to which it would be extremely difficult to attach any physical meaning. We may, therefore, conclude that if the other data are correct, then the supposed molecular weight must be in error (see Johnson and Rowe, 1961).

(3) From Equation (10.11), we may calculate an approximate value for the expected anhydrous molecular diameter, $2b$. It would be reasonable to assume $\bar{v} = 0{\cdot}73$, $a/b = 50$. The value yielded for the expected diameter ($2b = 66$ Å) differs so very much from the diameter observed in the electron microscope that it is clear that the two findings cannot be compatible. Probably an aggregation occurred during the drying down to prepare the specimen for electron microscopy.

(4) The ratio of the sedimentation coefficients is so nearly equal to $2^{\frac{2}{3}}$ (see answer to question 1 above) that it is highly probable that the enzyme dimerizes without significant change in shape or effective volume.

(5) A plot of the data shows a line with a slight positive curvature, the gradient increasing with x to the extent of about 10 per cent. The sample is therefore not completely homogeneous, but the data do not permit of any exact definition of the degree of inhomogeneity.

(6) The frictional ratio (f/f_0) can be calculated, combining Equations (7.1) and (7.3). The value yielded (0·96) is less than unity, suggesting that either the protein behaves anomalously, or the data are in error.

(7) The time required to reach equilibrium is inversely proportioned to the diffusion coefficient of the solute (Equation (6.11)); which for a given solute itself be inversely proportional to the solvent viscosity. For an aqueous solvent, the time required will therefore increase rather less than two-fold in going from 20°C to 0°C. The 'stability', however, increases four-fold with a 20°C decrease in temperature. Hence it would be advantageous to work at 0°C rather than at 20°C.

(8) To an approximation, we may calculate the ratio c_B^{β}/c_B^{α} (Equation (8.3)) by ignoring radial dilution, using the approximate Equation (8.4) to calculate s_B^{α}, and taking:

$$c_B^{\beta} = \tfrac{1}{2}(c_B^{\gamma} + c_A^{\gamma})$$

The ratio given is 1·4, from:

$$\frac{c_B^\beta}{c_B^\gamma}=\frac{(6{\cdot}80-0{\cdot}5\times3{\cdot}2)-(4{\cdot}15-0{\cdot}5\times2{\cdot}9)}{(6{\cdot}80-0{\cdot}5\times3{\cdot}2)-(4{\cdot}15-0{\cdot}25\times2{\cdot}9)}$$

This indicates that a very significant 'Johnston–Ogston' effect is present, and that the relative peak areas could not be used (without extensive correction) to give an estimate for the exact proportions of A and B in the mixture.

(9) The protein is bovine serum albumin. Values of about 68,000 are yielded. If values of the height of the trace (dc/dx) and of the gradient of the trace (d^2c/dx^2)—both relative to the solvent trace—are estimated at an x value close to the mid-point of the column, then the M value calculated can be referred to the initial solute concentration (Equations (6·8)–(6.10)), assuming that no significant quantity of heavy, aggregated solute material was present.

REFERENCES

Jeffrey, P. D. and M. J. Pont (1969) *Biochemistry*, **8,** 4597.
Kado, C. I. and D. R. Black (1968) *Virology*, **36,** 137.
Mueller, H. (1964) *J. biol. Chem.*, **239,** 797.
Riethel, F. J. and J. D. Sakura (1963) *J. phys. Chem.*, **67,** 2497.
Rowe, A. J. (1960) Ph.D. Thesis, University of Cambridge, England.
Rowe, H. J. and A. J. Rowe (1970) *In press.*
Schumaker, V. and P. Adams (1968) *Biochemistry*, **7,** 3422.
Svedberg, O. and F. Sjorgren (1929) Quoted in Alexander, A. E. and P. Johnson (1949) *Colloid Science* Table 11.1, Clarendon Press, Oxford.

Index

Actin 145
Activity and s value correlation 120
Adiabatic expansion of rotors 37, 122
Agar 112
Aggregation 99, 107, 138, 162
Air turbine 6, 12
Albumin 7, 65, 68, 98, 112, 118, 154, 166
Aldolase 152
Alignment of optics 23, 160
Allosteric effects 147, 149
Analytical ultracentrifuges 9
Anderson (zonal) rotor 12, 122, 124, 125, 134
Angle head rotor 12, 123
Angular velocity 2, 129
Antigen–antibody 111
Approach-to-equilibrium 23, 54, 75, 160, 161
Archibald method 23, 54, 75, 160, 161
Archimedes principle 3
Aspartate transcarbamylase 150
Asymmetry of particle, effect of 118, 144
Automatic integrator 130
Axial ratio of particle 90, 145, 150, 151

Bacterial extract 37
Balance, static and dynamic 123
Band capacity 135
  centrifugation 37
Band-forming cell 28, 131
Baseline irregularities 23
Beckman Model E ultracentrifuge 9
Beer–Lambert law 17, 97
Bence–Jones protein 163
Beta function 90, 92, 163, 165
Blood group substances 93
Buoyancy method for M 79
Buoyancy term, $1-\bar{v}\rho$ 3, 56, 58

Caesium salts 80, 118, 127, 131
Cahn electrobalance 58
Calibration cell 22, 98
Capacity of density gradient 134
Carbamyl phosphate 150
Carboxyhaemoglobin 7
Cell,
  band-forming 28
  calibration 22, 98
  double sector 28, 97, 152
  interference 25, 29
  multichannel 29, 68
  partition 30, 121
  single sector 28
  synthetic boundary 28, 29, 37, 47, 63, 70, 73, 162
  wedge 27, 28, 35, 160
Cellulose 112
Centrifugal field 1
Chain (sub-unit) molecular weights 151
Charge effects 32, 45, 61, 89, 151, 160
Charlwood's method for M 73, 162
Christ analytical ultracentrifuge 11
Chromatography 104, 112
Chymotrypsin 103, 109
Citrate synthase 151
Collision reflection 5
Complexes 109
Computer 17, 55, 156
Conformation 84, 146
Continuity equation 50, 110
Convection 32, 37
Counter-ions 33
Counterpoise (interference) 25
Creeth's steady state method 48
Cylindrical lens Schlieren optics 20

Denatured protein 93, 161
Density, determination 57, 59, 95, 120
Density, inversion 131
Density gradient, capacity 134
Density gradient, production 131
Depolymerization 154
Diffusion coefficient 3, 39
corrections 43
determination 44
related to *f* 40
Diphenylamine reaction 118
Dissociation constant 102
Distortion correction 65, 69, 98
DNA 61, 80, 92, 106, 125, 146
Donnan effect 34
Double-beam scanner 9, 17, 63, 66, 153
Double-sector cell 28, 97, 152
Droplet formation 135
Drugs 146
Dye-labelled protein 39
Dynamic balance 123
Dynamic equilibrium 103

Effectors 147, 150, 151
Electrobalance (Cahn) 58
Electronic speed control 9, 162
Electron microscopy 146, 150, 151, 163, 165
Electrophoresis 112, 138
Electroviscous effect 89
Ellipsoids, frictional ratio table 119
Equilibrium dialysis 152
isodensity 80, 122, 130
Equivalent time of centrifugation 52
Erythrocytes 118
Esterase 149
Ethidium bromide 148
Excluded volume 92, 148
Expansion factor 93

F-actin 145
Fick's laws of diffusion 39
Fingerprinting 138
Flow equation 51
Flowing junction diffusion cell 46
Fluid lines (zonal rotor) 127
Fluorochemical FC–43 68, 72, 161
Frictional constant 1, 40, 116, 119, 147
ratio 85, 119, 164

Fujita–MacCosham equations 55
Fujita solution for peak spreading 48

Gas constant 4
Gaussian peaks 11, 45, 135
Gelatin 112
Gel electrophoresis 138
Gels 112, 162
General ultracentrifuge equation 50
Gilbert theory 102
Gilbert–Jenkins theory 109
Glucose-6-phosphate dehydrogenase 107
Glycoproteins 93
Guanidine hydrochloride 94, 103, 139, 151, 152, 153

Haemocyanin 6
Haemoglobin 7, 154
Heavy water 59, 68
Helical molecules 146
High speed equilibrium method for *M* 67
Hinge point 66
Homogeneity criteria 137
Huggins equation 88
Hydrodynamically equivalent ellipsoid 90
Hydrogen bonding 102
Hypersharp peaks 32, 112

Ideal sedimentation 122
Influenza virus 127
Infrared sensor 9
Integration (trapezoidal) 77
Integrator, automatic 130
Intensity of Rayleigh fringes 26
Interactions 106, 154
Interference cell 25, 29
Interference effects on Schlieren peaks 22
Intrinsic viscosity 87, 94, 163
Isodensity methods 122, 130
Isokinetic gradient 116, 119
Isopiestic technique 152
Isopyknic method 122, 130

Johnston–Ogston effect 99, 112, 165

Kraemer's hydration correction 86

LaBar's method for M 71
Lactoglobulin 107
Lamm continuity equation 50, 52
  solutions 55
Lamm diffusion cell 46
  method for M 73, 141
  scale optical system 6, 44
Lansing and Kraemer method for M 70
Le Chatelier effect of pressure 107
Light scattering 61, 99, 150, 162
Lipoproteins 92, 98, 116, 121
Long molecules, width 145
Long molecules, swelling 146
Low speed equilibrium methods for M 63, 70
Lysozyme 134, 142

Maleic anhydride 108
Markers for s values 38
Mediated dimerization 106
Membrane proteins 92
Meniscus depletion method for M 67, 163
  position 36, 76, 161
Mercaptoethanol 94
Mercuric ions 163
Mercury pool contamination 159
Metal ions 154, 163
Methyl orange 154
Metrimpex analytical ultracentrifuge 12
Molecular anatomy program (MAN) 127
Molecular weight 60
  averages 61, 109, 141
Molecular weight determinations
  Archibald approach-to-equilibrium 75, 161
  buoyancy method 79
  chain (sub-units) 151
  Charlwood 73, 162
  LaBar 71
  Lamm 73, 141
  Lansing and Kraemer 70
  Mueller 79, 160
  Rowe and Rowe 141, 142, 164
  sedimentation–diffusion 65
  Trautman (Archibald) 76
  Van Holde and Baldwin 73, 141
  Yphantis meniscus depletion 67, 162, 163
  Yphantis and Roark 141
Monochromator 15, 17
MSE analytical ultracentrifuge 9
Mueller's method for M 79, 160
Multichannel centrepiece 29, 68
Myoglobin 154, 155
Myosin 79, 91, 112, 139

NADP 108, 151
Neurath diffusion cell 46
Newtonian behaviour 87
Non-equilibrium thermodynamics 41, 54, 60
Normalized data 79
Nucleic acids (also see DNA and RNA) 61
Number average molecular weight 61

Offset slits (interference) 24
Oil turbine drive 5
Oligosaccharide 161
Optical alignment 23, 160
Optical attachments for preparative ultracentrifuge 13
Optical systems 15
Orcinol reaction 118
Osmotic pressure 3, 40, 118

Partial specific volume 3, 56, 152, 159
Partition cell 30, 121
Pepsin 112
Performance index of preparative rotor 129
Perrin's equation 86, 145
Phaseplate 22
Philpot–Svensson optics 15, 20
Pig liver esterase 149
Planimeter 160
Plasticizer 72
Plateau region 52, 138
Plug flow 112
Poiseuille's law 95
Polyoma DNA 146
Polyvinyl pyrrolidone 118
Porous plug analogy 32
Preparative ultracentrifuges 12
Pressure dependent system 107

Probability integral 43, 47
Purity, criteria 137
Push–pull slit 23

Quantitative analysis 97

Radial dilution correction 36, 53, 77, 98, 102, 113
Radioactivity labelled protein 39
Raised baseline 23
Raster light source 27
Rate sedimentation 121
Rate-zonal method 121, 130
Rayleigh optics 15, 23, 63
Reduced viscosity 87
Refraction methods 15
Refractive index 17
Relaxation of fringes 71
Resolving power of centrifuge 5, 159
Ribonuclease 67, 72, 73, 142
Ribonucleotide reductase 163
Ribosomes 37
RNA 39, 125
Rotors, angle 12, 123
  and cell design 122
  performance 128–130
  stretching 37
  swing-out 123
Rowe and Rowe method for *M* 141, 142, 164
RTIC (rotor temperature indication and control) 9
Rubidium salts 118

Salt effect on viscosity 88
Sapphire windows 65, 69
Saucer-shaped baseline 23
Scanner 17, 63, 153
Schlieren optics 18
Scintillation mixture 118
Sedimentation, conformation effect 85
  –diffusion 65
  equilibrium 4, 53
  ideal conditions 122
  potential 33
Sedimentation coefficient 3, 31
  charge effects 32
  concentration effect 32
  corrections 34, 158
  determination 35, 38
  ionic strength effect 108
Selection of rotors for preparative work 127
Selective binding of ligands 152
Self-sharpening effect of peaks 32, 46
Skew peaks 23, 47
Slits, offset (interference) 24
  push–pull (light source) 23
  swinging (camera end) 67, 71
  symmetrical (interference) 26
Sodium dodecyl sulphate 93
Sol–gel transitions 112
Sophianopoulos–Van Holde equation 108
Specific refraction increment 17
  for sucrose 22, 98
  for proteins 98
Specific viscosity 87
Speed control and measurement 6, 9, 10, 12, 62, 162
Stabilizing device for rotor 123
Static balance 123
Staudinger index 87
Steady state method (Creeth) 48
Stoke's equation 1, 84
Strohmaier cell 38, 124
Sucrose, density and viscosity table 117
  refraction increment 98
Sulpholane 118
Supercentrifuge 2
Supercoiled DNA 146
Svedberg, equation 4
  unit 3, 31
Swinging bucket rotor 12, 123, 129, 133
Swinging slit 67, 71
Swirling effects 123, 124
Synthetic boundary cell 28, 29, 37, 47 63, 70, 73, 140, 160, 162

Teller's method for *M* 141
Temperature 9, 11, 12, 37, 62, 122, 159
Thermodynamic factor 4, 41, 60
Thermoelectric cooling 11
Theta solvent 76
Thorium oxide 118
Tilted baseline 23
Time to equilibrium 68

Titanium 8, 13, 127
Toepler–Schlieren optics 6, 10
Toolmaker's microscope 36
T_3 phage 134
Transient state 54, 160
Trapezoidal integration 77
Trautman plot 76, 79
Trimethyl phosphate 118
Tube puncturing 131
Tube slicer 131

Ultraviolet, recorder 125
  scanning optics 13, 15, 66, 153
  single-beam system 6, 17
Urea 70, 92

Van der Waals' forces 102
Van Holde and Baldwin method for *M* 73, 141
Van't Hoff equation 40
Viruses 39, 127, 162
Viscosity, and conformation 87
  coefficient definition 87
  increment 90
  measurement 94
Volume fraction 90

Wales–Van Holde ratio 93
Wall effects 118, 125
Watson–Crick double helix 147
Wedge cell 27, 28, 35, 160
Weight average 61
White light fringe 27, 66
Width of long molecules 145
Windows, sapphire 65, 69
  wedge 27
Wobble of rotors 13, 47, 123

x–y measurements 36

Yphantis meniscus depletion method for *M* 67, 162, 163
Yphantis and Roark method for *M* 141

z-Average 61, 73, 141, 163
Zeiss diffusion cell 46
Zeroth fringe 27, 66
Zero time correction, diffusion 45
  sedimentation 36
Zonal methods 130
  (Anderson) rotor 12, 122, 124, 125, 134
  runs 37
Zone capacity 135